NORMAN CURREY

AIRPLANE STORIES AND HISTORIES

AIRPLANE STORIES AND HISTORIES

N O R M A N   C U R R E Y

BOOKSIDE Press

BOOKSIDE Press

BookSide Press
877-741-8091
www.booksidepress.com
orders@booksidepress.com

CONTENTS

INTRODUCTION

My file "A" was always chock-a-block full until I reorganized it into sections such as aerodynamics, design, history, materials, structures, and so on. The contents of this book are taken from my "general" file – a collection of interesting and sometimes unusual accounts/ events in aviation.

My own career in aviation began in June 1941 when I joined the local squadron of the newly-formed Air Training Corps, and on the 29th of that month, I made my first flight – in a Vickers *Wellington* bomber from Driffield (Yorkshire) aerodrome. This occurred while our squadron attended one week of training with the Royal Air Force (RAF). During a similar week the following year, we were bombed while there! The following year I made my second flight – this time in an Airspeed *Oxford* twin-engine trainer. By this mid- 1943, I had decided to follow a career in aeronautical engineering.

After leaving high school, I was accepted at the de Havilland Aeronautical Technical School at Hatfield, just north of London. This provided a four- year college-level course in aircraft design, maintenance, and production engineering. I concentrated on design engineering, which entailed passing examinations in mathematics, materials, theory of structures, aerodynamics, and an all-encompassing subject called "design" that included subjects such as hydraulics, landing gear, controls, and so on. I mention this because it is different from how this subject is taught today. For example, we took some of the examinations at the Imperial College in London. In addition to the classroom work, we had to spend 20 to 30 hours per week working in various departments of the de Havilland Aircraft Company to become thoroughly familiar with the production and testing of aircraft. Prior to this, we spent several months in the school's workshops – operating machines of various types, sheet metal work, engines, and drawing office. During our time in the factory, we spent several months in each department, such as foundry, drop hammers, press shop, production line, flight test, instrument lab, aerodynamics, and stress office. Including workshops and classrooms, we worked about 10 hours per day for 5½ days per week and only had two weeks of summer vacation. I have read that this school was considered the best of its type anywhere, and I must admit that I have never seen one to equal it! A few years after I graduated, it became part of Hertfordshire University.

In today's aircraft industry, it is not unusual to work on only two or three aircraft in a lifetime. During the 1940s, the war spurred development and demand so quickly that we worked on several aircraft types within only a few years. In the six years at de Havilland, I worked on the DH *Mosquito* fighter-bomber (manufacturing and flight testing), DH *Vampire* twin-boom fighter (aerodynamics), DH 103 *Hornet* fighter (manufacturing), DH *Dove* small airliner (manufacturing), and the DH *Comet* airliner (structural analysis). I also obtained my pilot's license in a DH *Tiger Moth*.

After leaving de Havilland, I joined Avro Canada, located at the side of Toronto airport. For the next ten years, I helped design the CF-100 fighter, the *Jetliner*, and the CF-105 *Arrow* fighter. In my spare time, I was part of a very small design team that designed a single-engine bush plane, the Found Brothers FBA-2.

When Avro Canada closed down in 1959, I joined Lockheed in Marietta, Georgia, where I worked for 30 years. I spent most of my time in Preliminary/Advanced Design, where I worked on the C-130, *Jetstar*, C-141, C-5, and on many studies as well as Research & Development. Some studies included C-130s that were stretched considerably to carry passengers. We explored a variety of medium-size airliners (some at the request of Howard Hughes), some VTOL aircraft, a ground-attack aircraft, and an airplane that we called the NASA STOL (*Questol*). It was a competition from NASA to design an airplane that would be used to study many aspects of short take-off and landing, and it was a particularly interesting program to work on.

We won the competition and were ready to start detail design when the program was cancelled due to budget cuts. We examined versions of the C-130 to operate from aircraft carriers (and one did!). It was piloted by Lt. James H. Flatley when he repeatedly landed and took off again from the U.S.S. *Forestal* in moderate-to-rough seas in a KC-130F in 1963. Regular carrier operation requires a high sink-rate landing gear, and we did extensive studies on modifying the C- 130 gear to accomplish this.

During my time at Lockheed, I had the opportunity to present several lectures in the U.S. and abroad, wrote many technical articles and papers and one engineering textbook. Since then, I have done some consulting work and became a Fellow of the Royal Aeronautical Society. So that summarizes my credentials, and I hope that you'll enjoy reading the contents of File A.

CHAPTER 1
SIR GEORGE AND THE WRIGHTS

Sir George Cayley (1773 – 1857) has been called "The Father of Aerial Navigation," "Father of Aviation," and "The true inventor of the aeroplane…". However, he did not make the first flight in a powered airplane and could land on the ground equal to or more than the height from which it took off. That honor went to the brothers Orville and Wilbur Wright in 1903—46 years after Cayley died. He did provide the concepts and fundamental principles, though, including control of pitch, roll, and yaw. Still, in the mid-1880s, the internal combustion engine was not yet available, so he didn't have the means to put his ideas into practice.

Sir George was a Yorkshireman, born in Scarborough on Yorkshire's North Sea coast. He lived in Brompton (four miles from where I lived for the first 17 years of my life), a village some eight miles inland from Scarborough on the edge of the Yorkshire Moors. His mother, Isabella Cayley, was a very astute observer of nature and was aware of her son's peculiar talents. She encouraged him to observe the natural world and document his findings, particularly relating to birds. When he was a teenager, she sent him to be tutored in Nottingham. He met the daughter of the Rev. George Walker, one of his tutors who was a mathematician and mechanic and a Fellow of the Royal Society. His daughter was a beautiful redhead and noted for her bad temper tantrums. Notwithstanding that, he fell in love with her, and they married. It turned out to be a very successful marriage, and the villagers were somewhat amused, and some dismayed to see her riding horseback with legs astride the saddle and smoking a pipe!

Sir George studied bird flight in considerable detail and documented his work—their dimensions, wing area, weight, speed, heart-beat, and flaps per minute. He conducted and documented many experiments to define and elucidate the effects of various wing sections and other aspects of flight, including the movement of the center of pressure as it relates to wing angle and wing shape. Just as Beethoven, Mozart, and Chopin were "born" musicians; Captain James Cook, Marco Polo, and Leif Erikson were "born" explorers; Napoleon, the Duke of Wellington, and George Patton were "born" to lead armies; Sir George Cayley was a born inventor. Not only did he prepare the groundwork for powered flight, but in his "spare time," he invented the caterpillar tractor, used on the tanks that were invented for World War I. He also designed, made, and tested missiles with cruciform fins, similar to today's missiles. He fired these into the bay at Scarborough and demonstrated how the cannonball's range could be substantially increased by changing to a streamlined shape. He was also the first landowner to provide the laborers on his estate with an acre of land to cultivate for their own use. He did considerable work on the safety of trains, replacing the simple bumper at the front with compressed air shock absorbers and having automatic braking at the instant of impact. In yet another area, he became well-known for his expertise in land drainage. For example, in one area, he converted a large area that was essentially a swamp into one that grew crops of grain. In addition, he designed and made an artificial hand for one of his workers whose hand had to be amputated after an accident. It worked very well, even allowing the worker to write and pick up a nail from the floor!

It is interesting to note how researchers are once again studying bird flight and using morphing structures to emulate their aerodynamic efficiency in today's jet-powered world. From his studies of bird flight, Cayley paid particular attention to the effects of wing camber to provide gliding capability. But he didn't follow the idea of other inventors attempting to design a flapping wing (ornithopter) type of vehicle. Instead, he correctly pronounced that the best way for humans to fly was to sit in a vehicle with a fixed wing of appropriate design to develop sufficient lift by applying forward power, like a propeller powered by some kind of engine.

He designed and built a rotating arm, like a one-bladed helicopter blade, at the end of which he mounted a series of airfoils (miniature wings) of various cross-sections. He tested these at multiple angles to see how the various amounts of tilt—the angle of incidence—affected lift, from which he found the optimum angles for wings with various amounts of camber. His reports showed details of the wing sections, close to those used for many years to come and used by the Wright brothers.

He also used the rotating arm to explore the streamlined shapes used for airships. Balloons had been invented in 1783 when Cayley was just ten years old. He did not pay too much attention to them. Presumably, he considered them merely to be bags of hot air that could lift people from the ground and whose course depended upon the direction of the wind. He did, however, think seriously about airships and even went so far as to prepare preliminary designs. He realized that their size (425 feet long) was beyond his capabilities, but he provided fairly extensive details. He thought a hydrogen-filled airship would be too dangerous due to its fire hazard, so he contemplated using the tried-and-true method of hot air. He used a steam engine to heat it with a chimney going up into the bag. He recognized the need to minimize leakage for his hydrogen version by using one bag of thin-oiled skin inside another of coarser material. He suggested paddles for forward motion, and he had a rudder for direction control. In both of his airship designs, the passenger/crew module was a boat-like structure suspended beneath the bag. A final note on airships—the world had to wait until 1874 (17 years after Cayley's death) before any more serious work was done on this subject when Graf Ferdinand von Zeppelin began his work.

After a long series of rotating arm tests, he built a crude model, the first true airplane mode, to test his findings. It had an adjustable tail with a fin and horizontal stabilizer and a movable weight to adjust the center of gravity. I find the following excerpt from Cayley's notes on this model to be a typical example of his work, "at a velocity of 21 feet per second, it would support 5.6 ounces, whereas by the tables of circular velocities there appeared only a support of 1.5 ounces. Crows have a foot surface and move 34.5 feet per second, supporting a pound, and the angle is not perhaps more than 3 degrees or 4 degrees. However, their wing is concave, which resists much more than a plane surface…."

He became aware of the significance of wing aspect ratio, the long-span, narrow-chord wing has a high aspect ratio, and its camber, which Cayley called concave. Seagulls were of particular interest to him in this regard. To quote him again, "I am apt to think that the more concave the wing to a certain extent the more it gives support and that for slow flights a long thin wing is necessary, whereas for short quick flights a short, broad wing is better adapted with a constant flutter as the partridge and pheasant." In a nutshell, there is a difference between a long-range bomber and a short-range fighter! Cayley also considered some of the structural problems associated with the long thin wing. When he commented on a proposal for an airplane with such a wing supported by bracing wires attached to a kingpost mounted above the fuselage, he expressed serious doubts about its structural integrity. He thought it better to provide adequate lift by using a biplane or triplane arrangement. He even went so far as to prepare a design for a triplane.

Camber, however, was one of his principal contributions to aeronautics. He developed some wing sections from his studies, one of which was almost identical to the one used by the United States National Advisory Committee for Aeronautics (N.A.C.A), the forerunner of N.A.S.A., for its aerofoil 63A016-LB N-0016. It is amazing how precisely he could define the coordinates for the entire contour of this shape. It was based on the shape of a trout. A fish that he considered to have the best-streamlined shape! I have not been able to find any references to him ever pursuing the various degrees of camber and their dramatic effects on a lift, and I must assume that those explorations were left to following generations.

Before proceeding with the design of the complete airplane, he advocated the use of landing gear instead of a skid, so he became the first inventor of the tension wheel, similar to what we have on bicycles. He was always concerned about weight since each pound of it had to be compensated for by additional lift, a precious commodity at that time. So he replaced wood spokes with light metal spokes, and the rim became the primary structural item. He even had a simple device to adjust the tightness of the spokes. With such a gear, he could accomplish a takeoff run and a landing run—vital elements for an airplane to develop lift from airflow speeding over the wing.

Another item to be addressed was a power unit. The steam engine in vogue at that time was far too heavy. The internal combustion engine hadn't yet been invented. So, he examined other types but without success. Consequently, his quest for powered flight had to be abandoned, and he concentrated on the design of the vehicle itself.

Two of his vehicle designs interested me in particular. One was essentially a convertiplane and the other a true airplane in which flights were made. The convertiplane had two rotors on each side of the fuselage rotating in opposite directions to eliminate torque. These were arranged helicopter- like for takeoff, and he had twin pusher propellers for forward thrust. All of which were interconnected to be driven by a centrally-located engine. In Cayley's words, it was "capable of landing at any place capable of receiving it, and of ascending from that point, and capable of remaining stationary, or nearly so, in the air."

The second design was a follow-on from a series of glider experiments. In 1809, he successfully flew his first full-sized, unmanned glider with a 300 square-foot wing from a nearby hilltop. His detailed account of that event is the first in the history of a full-sized, properly-airplane flight. In 1849, built a full-sized triplane based on the premise that stacking wings would provide adequate lift in a more compact shape. He flew that glider successfully with a ten-year-old boy aboard. In that same year, he conducted flight tests of elevators and rudders. Then in 1852, he produced his manned monoplane. The design that I referred to as the second design that interested me. It had operable control surfaces, wing dihedral, a full tail, an elevator/rudder, and a tricycle landing gear. In the following year, he flew it with his coachman sitting in "the nacelle." It was not until 1908 that another airplane, powered or unpowered, was to fly with all of these features. The only thing it lacked, compared to airplanes of much later vintage, was roll control. A replica of that airplane was built and successfully flown in the 2003 event that celebrated the 100th anniversary of the Wright Brothers' first flight. After the 1853 flight, Sir George's granddaughter, a certain Mrs. Thompson, said, "Of course, everyone was out on the high east side and saw the start close to. The coachman went into the machine and landed on the west side at about the same level. I think it came down rather a shorter distance than expected. The coachman got himself clear, and when watchers had got across, he shouted, 'Please, Sir George, I wish to give notice. I was hired to drive and not to fly.' That's all I recollect."

In addition to his aeronautical activities and inventions, Cayley seems to have had boundless energy and general enthusiasm for life. He was active in politics, having been elected as a Whig Member of Parliament in 1832. He helped found the Yorkshire Philosophical Society, the British Association for the Advancement of Science, and the Regent Street Polytechnic Institution. He died peacefully at Brompton Hall on December 15, 1857— twelve days before his 84th birthday.

Wilbur Wright said (in 1909): "*About 100 years ago, an Englishman, Sir George Cayley, carried the science of flying from a point which it had never reached before and which it scarcely reached again during the last century.*"

Other pioneers made valiant attempts to contribute to aeronautical science after Cayley's death, and some died in the process. In summary, all of Cayley's followers were "glider guys" – all of them hanging beneath gliders and maneuvering by moving their bodies (similar to the hang-gliders of today). Otto Lilienthal (Germany) (1848-1896) was probably the best. He was followed by Percy Pilcher (1866-1899) in the U.K. and Octave Chanute in the U.S. In France, Clement Ader built a monoplane powered by a steam engine that was unmanned and had no control. It managed to fly for about 160 feet before it crash-landed.

Professor Samuel Pierpont Langley came close to beating the Wright brothers for the top honors when he attempted to fly his machine, called The Great Aerodrome, three months before them. The ensuing arguments, accusations, and acrimony disturbed the serene world of aeronautics for many years. Langley was Secretary of the prestigious Smithsonian Institution who conducted experiments relating to airplane flight since 1866, including two models powered by steam engines that flew for distances up to ¾ mile in 1896. In 1903 he put his ideas into practice by building a full-scale airplane. However, it crashed during its launch. The arguments that followed centered upon the assertion, by various "experts," that the fault did not lie with the airplane but in the launching apparatus. It certainly would have flown, thereby becoming the first powered controllable airplane to fly. Langley died in 1906, so he was not involved in the arguments, but the Smithsonian was convinced of his flight by inaccurate information supplied to them. It sullied the reputation of that august body. In retrospect, if a prototype crashes on takeoff, it is not considered a success in today's world! It is comparable to a soccer team saying they'd have won the game if the opposition's goalposts had been a little further apart!

The basis for the expert's claim was that Langley's airplane was later proved to be flyable due to tests conducted in Hammondsport in 1914. Fellow aviation pioneer Glenn Curtiss directed the reassembly and modifications to Langley's plane and flew it from Lake Keuka in New York State. One statement said that all he had done was install

7

floats and the necessary trussing for them, which was later found to be a gross understatement. The Curtiss team had not attempted to fly the original Langley machine, and they made many changes before achieving success. They increased the wing area and increased the engine power substantially. The Smithsonian seemed to be unaware of the magnitude of these changes for a long time. Still, in 1942 they were finally able to admit that Langley's airplane, in its original form, was not considered to be the winner of that battle for the "first airplane to fly." The restored Langley Aerodrome is now on display at the Smithsonian in Washington D.C. The label on it has now been changed to delete any statement implying that it was the first heavier-than-air craft capable of sustained flight under its own power.

And now we come to **Wilbur and Orville Wright.** Wilbur was born in April 1867 and Orville in August 1871. Their father was Bishop Milton Wright of the United Brethren Church, a somewhat dour fellow who encouraged them to pursue their inherent interest in mechanics. This was extended by their mother, a born tinkerer. They lived in a clapboard house in Dayton, Ohio, had no university training, and in 1892 they decided to start a bicycle shop, The Wright Cycle Co., the first to sell them and then, three years later, to sell their own make. One was called the Van Clive and another the Wright Special.

They began to show an interest in f lying when they read about Lilienthal being killed in a gliding accident in 1896. They contacted the Smithsonian, and in reply, they received a list of literature that they could study and several pamphlets. Among the latter were Professor Langley's *Story of Experiments in Mechanical Flight* and some accounts of Lilienthal's work. From there, they realized that powered flight had been studied for many years by some of the brightest scientific minds, dating back as far as Leonardo da Vinci. Still, despite this, the two humble bicycle-makers from Dayton decided to see what they could do. They received encouragement from Octave Chanute, who had done some exploratory work in this area and a past president of the American Society of Civil Engineers.

They thought they could design a structure with adequate strength and minimum weight, and thought a suitable internal combustion engine could be built. The one big item they wanted to solve first was control. The lack of which was causing deaths among other experimenters. From observing pigeons carefully, they concluded that the wing tips needed to be independently twisted by the pilot, known as wing warping), to provide roll control. They decided that "a small movable horizontal surface in front of the main planes" would do the trick for pitch control. In today's language, it's called a canard, and even the Concorde had one!

In late 1899, during the off-season for bike sales, they built a five-foot span biplane kite with warpable wing tips. These could be controlled from the ground, showing that their system worked. So, they went to the next step—building a larger kite with a man on board, which they flew from Kitty Hawk, North Carolina. From a series of incremental tests, they evaluated the effects of incidence angle, speed, and pitch control. They were dissatisfied, however, with their wing warping, and they came close to disaster several times, so, as they say nowadays, back to the drawing board! They built a larger glider and began testing it at Kitty Hawk in July 1901, and after some modifications to the wing camber, it glided well. However, they were still not satisfied with the lift/drag ratio, so they decided to build a crude wind tunnel. Probably, the world's first. Using this, they examined a series of more than 200 airfoils in two months, and they subsequently tested their choice section on a large glider in 1902. As part of these experiments, they were able to define, for the first time, the actual coefficients to be used to determine the actual lift and drag forces.

They made some modifications to the 1902 glider, increasing its aspect ratio and wing area, the front stabilizer's area, and the area of the two vertical fins. The total weight, including the pilot, was about 265 pounds. With Orville as the pilot, all went well until he reached a stall angle, and as expected, this had dramatic consequences. From a height of about 30 feet, it dropped to the ground. Orville escaped uninjured but wiser! They were aware of what had happened and made some changes to the design. They also replaced the two vertical fins with a single movable surface (i.e., a rudder) and connected it with the wing- warping control.

By the spring of 1903, they faced the two remaining problems: finding or developing an engine and a propeller. They tried unsuccessfully to find an engine that met their requirements, so they decided to make one! It needed to have at least eight horsepower and weigh no more than 200 pounds. The one they built was just right – it developed 16 hp for a few seconds and then 12 hp. It weighed 170 pounds. Not bad for amateurs!

For the propeller, they studied marine propellers and then concluded that it was merely a wing moving in a spiral track, but the speed of each section increased as it approached the tip. They also realized that the stationary thrust achieved would be different as the airplane moved forward. They had no way of knowing how that would work out, so they decided to make one and fly it. They also decided to have two propellers on their machine, rotating in opposite directions to prevent any gyroscopic effects.

The airplane they built was similar to their 1902 glider, but it had the motor located beside the pilot and the propellers chain-driven from the engine. They'd increased the engine horsepower, wing area went up to 510 square feet, and total weight approached 800 pounds.

At the end of September 1903, they returned to Kitty Hawk to find that storms had severely damaged their building, and you really have to visualize the scene. Kitty Hawk is on a large sandbar off the Atlantic coast, a bleak expanse of land with relatively little vegetation constantly battered by wind and waves. So, task number one was to repair their building. They were then delayed for several weeks by more bad weather. Then on December 14, Wilbur climbed aboard and made the first attempt to fly their machine. It rose so quickly and sharply that it stalled, resulting in a broken skid that took two days to repair. Following more bad weather, they decided to try again on the 17th. There had been a bitterly cold wind the previous night, and ice coated the various pools. Wind speed was up to 27 mph that morning, and the nearby Kill Devil Hill Life Saving Station was alerted that they were about to fly—this was to notify neighbors of the attempt. Five people showed up to see the event, and none represented the press.

The engine was cranked up, and they let it run for a few minutes to warm up. Then, with Orville as the pilot and Wilbur steadying the wing tips, it was launched from its greased track. After a takeoff run of less than 40 feet, it rose slowly into the wind and flew for twelve seconds. At this point, one of the watchers, Johnny Moore, ran across the dunes shouting: "They done it! They done it! Damn'd if they ain't flew!"

Wilbur followed with an eleven-second flight, and within an hour of the first flight, Orville went up again for fifteen seconds. Then to wrap things up for the day, Wilbur took off in the early afternoon and flew it for 59 seconds, covering a distance of 852 feet.

THIS WAS THE BEGINNING OF FLIGHT AS WE KNOW IT – THE FIRST SUSTAINED FLIGHT OF A CONTROLLABLE POWERED VEHICLE CARRYING A HUMAN BEING.

In January 1904, they began building a new machine with a new engine, and they started testing it at Huffman Prairie, a field six miles east of Dayton. In that year, they made more than 100 flights, including a complete circle, and flew distances up to three miles. The small field size necessitated them designing an assisted takeoff device that consisted of a derrick. A 1600-pound weight was dropped from 161 feet to exert a force through ropes and pulleys to pull the aircraft forward, enabling takeoff within 50 feet.

It is interesting to note that the American press did not pay too much attention to the Wright Brothers' activities for reasons known only to themselves. This contrasts greatly with newspapers in the U.K. and Europe, where they were hailed as heroes. The Aeronautical Society of Great Britain (later, the Royal Aeronautical Society), founded way back in 1866, printed in January 1904 a complete statement written by the Wrights describing their first flight in great detail. Major Baden-Powell, President of that Society, presented details in a lecture in March of that year. He was among the society members who also corresponded with Octave Chanute, a keen supporter and adviser to the Wrights. By 1906, the achievements of the Wright Brothers were becoming known worldwide. They concluded that their work was done and decided to concentrate on developing a better engine. They said their new engine, which weighed 160 pounds and produced 25 to 30 horsepower, would be capable of flying two men, with enough fuel, for distances of several hundred miles.

They responded to a specification from the U.S. War Department in 1907 for an airplane—the first such contract ever to be awarded. The Wrights were awarded $5,000 to build it. It was to carry two people (totaling 350 pounds) with enough fuel for a 125 miles flight at 36 miles per hour. Up to that point, they had received no funds to help them with their work on airplane development, so this must have encouraged them a great deal. They delivered the airplane, and Orville demonstrated it on September 19, 1908, much to the delight of Washingtonians, including the President and members of the Congress who flocked to see it. It was amusing to read that Wilbur and Orville would not be fazed by the importance of their audience in that they frequently kept them waiting until they were "good and ready to fly," whenever that was!

By 1908, Wilbur demonstrated their airplane at the Le Mans race track in France. In a letter to Orville, who was back in the U.S., he wrote, "The excitement caused by the short flights I have made is almost beyond comprehension. The French have simply become wild." Several members of Britain's Aeronautical Society went over to Le Mans for the event, and some flew with Wilbur. One of them, Griffith Brewer, was the first, and he said afterwards, "I enjoyed the greatest thrill it had ever been my lot to experience." The second member to fly was the Hon. C.S. Rolls (later the first half of Rolls-Royce). Major Baden- Powell stayed for a week and made two flights. In November of that year, the Aeronautical Society approved the presentation of a specially- designed gold medal to the Wright Brothers "in recognition of their distinguished services to aeronautical science." Griffith Brewer and Aeronautical Society Council-member Alec Ogilvie developed a deep and lasting friendship with the Wright family. There were many letters between them discussing their overall viewpoints on aeronautics, one of which was from Wilbur to Ogilvie telling him that the Wrights had arranged with the Short Brothers to make six Wright machines powered by engines from Messrs Bollee of France.

Wilbur started a f lying school at Pau, France, while Orville was recuperating from a serious accident at Fort Myer. With his sister, Katharine, Orville joined Wilbur there in March 1909. They developed a new control column that was the beginning of the "joystick" for many years. The stick was moved left or right for lateral control, and for pitch control, it was moved fore and aft. In October of that year, Orville visited the Short Brothers plant at Shell Beach, England, to check the progress being made in building their six airplanes, and as a result, his friend Ogilvie made his first powered flight in one of them.

Despite their previous assertion that they'd build no more airplanes, they did. In October 1910, they completed the Wright Racer, a smaller machine with a standard 4-cylinder 35-hp engine, and a wing span of 26 feet and 8 inches. It could fly at 56 mph and climb at 600 feet per minute. A later model with a higher-powered engine entered the Gordon Bennett race but crashed. Quite a few more aviators were active by this time, and Graham White won it in a Bleriot.

New pioneers appeared during all of these efforts by the Wright Brothers. For example, J.A.D. McCurdy became the first person in the British Empire to fly, using the Silver Dart, a machine of his own design in December 1908. Alexander Graham Bell, the telephone inventor, formed the Aerial Experiment Association in Baddeck, Nova Scotia, Canada. Along with fellow Canadians McCurdy of Baddeck, F.W. Baldwin, an engineer from Toronto, and Glenn Curtiss from the U.S. Curtis invented the aileron for roll control instead of using wing warping. In the U.K., France, and Germany, there were people such as A.V. Roe, Louis Bleriot, Henri Farman, Anthony Fokker, Geoffrey de Havilland, and T.O.M. Sopwith.

Wilbur died on May 30, 1912, from typhoid fever. Orville continued for a while, but World War I erupted, dramatically and quickly changing airplane design. So, the final part of the Wright Brothers story addresses the location of their original machine. In letters between Orville and the Aeronautical Society of Great Britain, there is some discourse. In reply to one letter, Orville said:

"Of course, the machine ought to be in the National Museum in Washington. But when one considers the way the officials of that Institution allowed the original Langley machine to be taken out of the Museum and the original materials of its structure to be destroyed by private parties to a patent litigation, so that the machine now hanging in the museum is not the original but is mostly a new machine with many of the restored parts of different construction from the original; and when one further considers that they have hung upon this machine a misleading label true neither of the new nor the original machine, one can have little faith that our machine would be any safer from mutilation or that it would receive a label from the present officers of the Institution any more truthful than that of the Langley machine.

"If I were to receive a proposition from the Kensington Museum offering to provide our 1903 machine a permanent home in the Museum, I would accept the offer, with the understanding, however, that I would have the right to withdraw it at any time after five years if some suitable place for its exhibition in America presents itself."

And so it was that the original Wright Flyer went to the Science Museum in London where it stayed until 1948, at which time negotiations with the Smithsonian led to it being moved to Washington. Before being moved, the Science Museum arranged with the De Havilland Aeronautical Technical School at Hatfield to build a replica. Work started in March 1946, and the instructor in charge was appropriately named Horace WRIGHT! Other instructors and students of the school used the nineteen sheets of drawings provided by the museum to create an exact duplicate. The propeller, for example, was carved out of a solid piece of wood, and the chains were hand-made to be precise copies of the original. They even went so far as to prepare the wood to have an appropriately antique surface. The finished restoration was handed over to the Museum in late 1948.

Sir George Cayley (1773-1857)

Wright Biplane

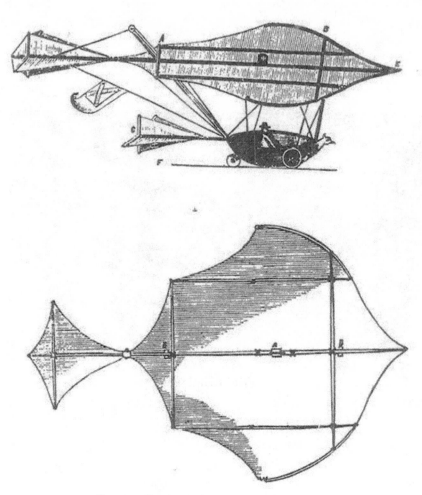

George Cayley's Governable Parachute

CHAPTER 2
OVERVIEW

A Heinkel He 111 bomber flew in from the coast, circled above us, and turned back towards the coast, presumably reconnoitering the radar towers on Staxton Hill. That was the first monoplane I'd ever seen! A few moments later, the second monoplane I'd ever seen, a *Spitfire*, zoomed in from the north and shot down the Heinkel. It was in the fall of 1939, shortly after Great Britain declared war on Germany.

Biplanes to Monoplanes

Monoplanes had been few and far between up to that time; in fact, airplanes of any type were not too common. In my part of Yorkshire, I saw the occasional DH *Tiger Moth* fly over from a nearby flying club. At Croydon, London's airport at that time, I saw the ungainly Handley Page Type 42 and Short L17 airliners— all of them biplanes. Some monoplane racing aircraft began to appear in various countries. In the mid-1930s, the first monoplane airliners showed up. Among these were the Ford Trimotor and the Boeing 247D, quickly followed by the Douglas DC-2/DC-3 series. In Britain, de Havilland produced the *Flamingo* and then the super-streamlined *Albatross*, but World War II stopped producing that aircraft. The Germans were a little more astute in that their monoplane airliner, the Junkers Ju52, was designed to be well suited for both commercial and military use. Hence, its production was unaffected by the upcoming war.

Monoplane flying boats made their first appearance in the mid-1930s and played a significant role in developing airline passenger service in the far corners of the world. Great Britain and the United States took the lead. In Britain, Short Brothers produced their *Empire* flying boats, and Imperial Airways used them to fly passengers south to Egypt and the Far East or South Africa. Juan Trippe, head of Pan American World Airways, was the prime mover of flying boat operations in the U.S. He started with the Martin 140 *China Clipper* carrying 26 passengers across the Pacific in 1935. He replaced these with the Boeing 314, a larger aircraft. In fact, it was large enough for passengers to have their meals in a restaurant-like setting with cloth-covered tables placed here and there and no visible seat belts! It is interesting to look back at how air travel has changed. First, everyone was dressed in their "Sunday clothes." They were fairly wealthy. A round trip flight from San Francisco to Manila would cost $1438, the equivalent of a typical year's wages at that time. Passengers were accommodated in relatively spacious cabins, and in some cases, full-size bunks were provided for sleeping.

On the military side, monoplanes were taking over, and by 1939 the Luftwaffe was leaping ahead with new bombers—Dornier Do 17, Heinkel He 111, Junkers Ju 86 about to be replaced by the Ju 88, and the Ju 87—and fighters, primarily the Messerschmitt Me 109, at that time. The Royal Air Force had the Hawker *Hurricane* (1935) and the just-in-time Supermarine *Spitfire*. It had also modernized its bomber fleet of Handley Page *Harrows* (1934) and *Hampdens* (1936) with Vickers *Wellingtons*, Bristol *Blenheims*, and Armstrong Whitworth *Whitleys*. In the United States, the old P-12 biplane fighters were replaced with Curtiss P-40 *Tomahawks* of "Flying Tigers" fame (with shark's teeth on their engine cowlings) and aggressive new Boeing B-17 *Flying Fortress*, Consolidated-Vultee B-24 *Liberator*, North American B-25 *Mitchell, and* Lockheed A-28/29 *Hudson*s, plus the U.S. Navy's Douglas A-24 *Dauntless*, Curtiss A- 25 *Helldiver* bombers, and Grumman's F4F *Wildcat*, and F6F *Hellcat* fighters.

It must be emphasized that the aircraft mentioned above are merely the significant/unusual/well-known ones of that era. I don't want, in any way, to diminish the importance of many others types not mentioned. For example, in the RAF, the Boulton-Paul *Defiant* fighter (1938) was quite a surprise to the Luftwaffe in 1940 due to its turreted machine guns behind the pilot. They shot down 37 enemy aircraft without losing them in their first appearance during the Dunkirk evacuation. However, their success was short-lived due to inadequate performance, maneuverability, and the absence of forward-firing guns. The latter shortcoming enabled a Me 109 to attack it head-on successfully! The Fairey *Swordfish* (1934) is an exception to most of the others in that it was a biplane and only had a top speed of 135 mph, yet it seemed to be remarkably suited as a torpedo bomber. They devastated the Italian fleet in 1940 and flew low-and-slow to torpedo the *Bismarck* battleship. Then there were the Gloster *Gladiator* fighters, the last RAF biplanes. Three of them (*"Faith," "Hope," and "Charity"*) were the only aircraft available to defend Malta in Mid-June 1940, and they fought well against superior odds.

In the U.S., many aircraft deserved "honorable mention." The Bell P 39 *Airacobra* (1939) had its engine behind the pilot, with a long extension shaft below the pilot connecting it to the propeller. It was used by the U.S. Army, RAF, and Russian air forces. The training aircraft produced by North American are very familiar to many older pilots. They were used by the U.S. Army as BT-9 and by RAF as the *Harvard*. The DC-3 is already mentioned, but in its military uniform, it became the famous C-47 *Dakota*—the most widely used air transport and air-dropper in World War II. The Martin A-22 *Maryland* was used by the British Fleet Air Arm in the early '40s for high-speed reconnaissance missions, and the Martin B-26 *Marauder* bomber was widely used by the USAAF and the RAF in North Africa, Italy, western Europe, and in the Pacific battles as the U.S. Navy M-1. The Lockheed A-28/29 *Hudson* has an interesting story. The British bought 250 of them right after the war started in 1939 and later ordered many more. It was a militarized version of the Lockheed 14 transport, complete with an upper fuselage gun turret and glass bomb aimer's nose. It attacked on German battleships anchored in Norway and conducted coastal command duties.

Battle of Britain

By mid-1940, the German army had overrun the Netherlands and Belgium and had outflanked the Maginot Line, allowing them to occupy France. This forced the British Expeditionary Force to evacuate from Dunkirk and then the Battle of Britain began. The Ju 87 dive bomber, which had proved itself to be highly effective, was no match for the *Spitfires* and *Hurricanes* and was soon taken out of the battle. Both RAF fighters were highly successful against the Do 17 and Heinkel 111, but when the Luftwaffe realized that, they started to provide fighter escorts, using the Me 109. However, the Me 109s only had a short range, about 80 minutes, limiting their usage and resulted in many of them not returning to base due to lack of fuel. The *Spitfire* was their most effective adversary. Its speed was comparable to the Me 109, and it had a tighter turning radius. Its downside was its armament and engine cutting out in a high-speed dive. Messerschmitt developed the twin-engine Me 110 in an attempt to provide a longer-range fighter, but it didn't work out too well. Its speed was less than the *Spitfire's*, and it was larger—an easier target to hit!

When Luftwaffe's chief, Hermann Goring, saw that the results of his failed attempts to destroy the RAF were not succeeding, he abandoned his large fighter-escorted daylight raids and switched to night raids on major cities, particularly London. Several books have been written about this phase of the war—Dunkirk, German plans to quickly end the war, invasion preparation, air wars, and the aftermath. I suggest that the reader examine such books as Time-Life Books' "Battle of Britain," an excellent and well-illustrated account of that interesting time.

The Ju 88 medium bomber appeared on the scene—a highly effective aircraft used typically in fast, individual, low-level attacks. Great Britain was building aircraft faster than Germany, and its fighter force was increasing quickly, much to the surprise of Herr Goring. Also, the RAF began to build up its bomber force. The Luftwaffe had been ordered to refrain from bombing London, but a navigational error caused some bombs to be dropped on the city during a raid on its dock area one night. When this happened, Churchill ordered an attack on Berlin using *Hampden* bombers. It is an excellent example of a relatively small (and accidental) incident having severe and unintended consequences. Both sides began bombing campaigns, with London being the main target. Huge raids ensued in the city. In the first major raid, there were about 300 bombers and 600 fighters in the Luftwaffe forces. Later in the war, the RAF responded with 1000-bomber raids.

The RAF introduced Avro *Lancaster,* Handley Page *Halifax,* and (to some extent) Short *Sterling* 4-engined bombers, and then the de Havilland *Mosquito* night fighter/bomber. These state-of-the-art aircraft were used effectively to attack landing barges along the French coast, industrial complexes in the Ruhr valley, Hamburg and Berlin. The *Mosquito* is a particularly interesting aircraft, and there will be more discussion of it later in this compendium. First, it was made of wood! Second, it was faster than most German fighters, even when it was being used as a bomber, and third, it was an exceptional design that enabled it to be very effectively used in a wide variety of roles. Its radar allowed it to shoot down many attackers at night, and in its high-speed pin-point bomber role, it swept in on unsuspecting targets to wreak havoc.

Britain's fighters were being upgraded. The *Hurricanes* were retired and replaced by the Hawker *Typhoon,* and the *Spitfire* underwent a succession of upgrades. All of these have top speeds of more than 400 mph, were very maneuverable, and were armed with multiple cannons plus, in some cases, machine guns.

The U.S. Enters the War

The United States entered the war after the Pearl Harbor attack in December 1941. This brought the B-17 *Flying Fortress* and the

B-24 *Liberator* bombers to the European theater, plus some fighters such as the Republic P-35 *Thunderbolt*, to act as escorts. Unlike the British bombers, the U.S. bombers were more suited to daylight operations; they sacrificed bomb loads for gun turrets, requiring more crew members. For example, the British *Lancaster* could carry up to 18,000 pounds of bombs but only had three turrets for its protection. In comparison, the *Flying Fortress* could only carry 6,000 pounds normally (or 12,000 pounds maximum) but was well-protected by gunners on the nose, tail, top, bottom, and sides. When in appropriate formations, the crossfire between gunners presented a formidable opposition to German fighters. However, losses were still substantial, and fighter escorts became paramount. The *Hurricanes* and *Spitfires* had an inadequate range for long-range bombing missions. So the P-47 *Thunderbolts* and Lockheed P-38 *Lightnings* were brought in, later augmented by the faster, newer, and more heavily-armed North American P-51 *Mustang*. This aircraft, which was flown by both the British and American forces, had four 20 mm cannons in its wings, and its use as an escort fighter resulted in daylight raids being very successful for the first time.

Two more aircraft were significant in the latter part of the war in Europe, and both were German fighters. The Focke-Wulf Fw 190, which entered service in late 1941, and the Messerschmitt Me 262, which entered service in July 1942. Another Messerschmitt aircraft, the Me 263, also deserves mention not because it was a great threat to the American/British bombers. But because it was the world's first rocket- propelled manned interceptor; however, its usefulness was hampered by its short 12-minute duration. It was also the world's fastest aircraft in 1944, with a top speed of 596 mph. The Me 262 was the world's first jet-powered aircraft to go into combat, and while it was particularly effective, it was grounded for lengthy periods due to problems with the materials used in its two engines. I have purposely omitted reference to the German V-1 Buzz Bomb. It was, by definition, I suppose, an aircraft, and it was significant, but it was essentially a flying bomb, shot off the ground from rails, that plummeted down to earth when its engine stopped. In Britain, the RAF's last piston-engine fighter-bomber was the DH 103 *Hornet*. A 470-mph aircraft with two Rolls-Royce *Merlin* engines that packed quite a punch—four 20 mm cannons in the nose and underwing racks that could carry two 1,000-pound bombs or eight rockets. However, it was too late for World War II.

In the Pacific area, the U.S. Navy aircraft predominated during the early period when the U.S. was island-hopping across to Japan. In mid-1942, there was the Coral Sea, then Midway, and in these, the objective was basically to destroy the Japanese navy, particularly its carrier force. It did this very successfully by using an aircraft from the small U.S. carrier fleet. Major Japanese forces were very surprised by the devastation caused by Douglas SBD *Dauntless* dive bombers, Douglas *Devastator* TBD torpedo bombers, Grumman TBD *Avengers* torpedo bombers, and Grumman F4F *Wildcat* fighters. Later in the Pacific War, we saw Chance-Vought F4U *Corsair* and Grumman F6F *Hellcat* fighters making a name for themselves.

In the land war in the Pacific theater, C-47s were flying "over the hump" to supply the Chinese, and inevitably there were the B-24s and B-17s bombing Japan. But one operation needs special mention not because it caused a great deal of damage, but because of its psychological effect. It was the Doolittle raid in Tokyo. Using North American B-25 *Mitchell* bombers that had never been designed for carrier operations, Lt. Col. James Doolittle led 16 of them, taking off from the U.S.S. *Hornet* soon after the Pearl Harbor attack and dropped bombs on Tokyo. This attack made the Japanese realize that they were not invulnerable. Consequently, they decided to occupy Midway, hence the ensuing battle that was considered by many to be the turning point in the Pacific war. This emphasizes the growing realization that air superiority was vital in modern warfare.

The Boeing B-29 *Superfortress* first flew in September 1942, and in June 1944, it made the first land-based bomber attack in the Japanese mainland. A little more than a year later, on August 6, 1945, a B-29 left Tinian Island and dropped the newly-developed 9,000-pound atom bomb (*Little Boy*) on Hiroshima. After a repeat performance above Nagasaki, these two knockout punches ended World War II. This removed any doubt that airpower could be decisive in war. One bomb had destroyed an entire city and saved many months of ground warfare.

The Jet Age

The "Jet Age" was now upon us. The first patent for a jet engine was issued to Sir Frank Whittle in January 1930. He had prepared a thesis on a new power plant while a cadet at the Royal Air Force College in 1928. Later, when he was Pilot Officer and later RAF Squadron Leader, he was so convinced in the concept of jet propulsion that he solicited private funds and formed a company called Power Jets. He designed an engine and got British Thomson-Houston (BTH) to build it. It ran for the first time on April 12, 1937, and was used to power the Gloster E28/39, an airplane especially built for it. That airplane made its first flight on May 15, 1941. The first jet-powered aircraft in the U.K.

The Germans were more energetic and had quickly realized the advantages of jet propulsion. Pabst von Ohain built an engine similar to Whittle's, Its first run was four months after Whittle's, but he had it installed in an aircraft and flown by August 27, 1939, about two years before Whittle's flight. It was flown in a Heinkel 178. And then for practical usage, in the Heinkel He 280, the world's first jet fighter.

The British and Germans jockeyed for supremacy in this area, and de Havilland flew their DH100 *Vampire* in September 1941, with their own DH *Goblin* engine. It became operational in April 1946. Gloster produced the *Meteor* twin-jet fighter in 1943 and the *Javelin* in 1951. Back at de Havilland, they followed up the *Vampire* with the *Venom* and *Vixen*. Six *Vampires* made the first flight over the Atlantic by jet-powered aircraft in July 1948. Unlike the Germans, the British fighters were not generally operational until the European War was over. The *Meteor* did see some minimal service against the V-1 Buzz Bomb.

The DH 108 *Swallow*, an all-wing research aircraft, was flying at a speed close to Mach 1.0 when unpredicted aerodynamic forces caused violent pitching oscillations that broke it apart, killing test pilot Geoffrey de Havilland Jr. This indicated the problems about to be addressed on both sides of the Atlantic. The German and Japanese aircraft industries were virtually out of the picture in the immediate post-war period, and the United States was playing catch-up in jet propulsion. Engine companies General Electric, Pratt & Whitney, and to some extent, Westinghouse started with versions of British engines and then developed engines of their own design.

The first U.S. operational jet aircraft was the Lockheed P-80 (later F-80) Shooting *Star* fighter designed by a hand-picked team working 60-hour weeks. Work started on June 24, 1943, and its first flight on January 8, 1944! It was powered by a de Havilland *Goblin* engine and later by a similar General Electric 133, and the USAAF immediately ordered 5000 of them. Two *Shooting Stars* were sent to England in early 1945 but didn't get into combat. The "C" model was introduced in June 1950, during the Korean War. At this point, it fought against the Russian MiG-15. The world's first dogfight between jet fighters occurred on November 8, 1950, and F-80s shot down a total of 37 enemy aircraft during that war for a loss of 14 of their aircraft.

The F-80 was succeeded by America's first swept-wing fighter, the North American F-86 *Sabre Jet*. It was highly successful in its first major operation— the Korean War. And was a worthy opponent to the MiG-15. They shot down about 850 enemy aircraft to lose only 56 *Sabres*. It first flew in October 1947, and early versions reached 680 mph. In a shallow dive, it could exceed Mach 1.0. It was succeeded by the F-100 Super Sabre, and this exceeded Mach 1.0 in its first flight, the world's first operational fighter to fly supersonically in level flight.

Supersonic Flight Begins

While North America and Lockheed were studying fighters to fly supersonically, USAF test pilot Charles Yeager climbed aboard a Bell X-1 to actually fly through the so-called "Sonic Barrier." It was a relatively small rocket-propelled aircraft and was dropped from the belly of a modified B-29 bomber. In the first few flights, he noticed the severe buffeting that Geoffrey de Havilland had encountered, but, in this case, the aircraft did not break up! The buffeting was caused by the shock waves formed as the air was compressed at near sonic speeds. The speed of sound is 764 mph at sea level, dropping to 660 mph at 36,000 feet, and the pressure waves that normally "warn" the air ahead of a wing to separate and flow smoothly over the wing cannot separate in time. Consequently, a shock wave of highly compressed air is formed, severely disrupting flow over the wing. Yeager decided to apply full power, and sure enough, the X-1 shot through the "barrier" and found everything smooth again on "the other side."

McDonnell developed the FH-1 twin-jet fighter for the U.S. Navy, followed by the F2H *Banshee*, the F-101 *Voodoo*, and the very popular F-4 *Phantom* for the USAF, U.S.N., and Marine Corps, plus various foreign air forces. Lockheed produced the F-104 *Starfighter*, and the "A" model flew in February 1957. Contrary to thinking at that time, it had a straight wing, but it was ultra-thin. It flew at a record speed of 1,405 mph, close to Mach 2, and was able to fly at very high altitudes. One of them reached 103,396 feet—another record! Northrop produced the F-5 in 1960 for the USAF, which was exported to many countries.

In Canada, a new company, A.V. Roe Canada Ltd., later Avro Canada, had been formed. A few key engineers were native Canadians, but most engineers were from the U.K. and U.S.A. They got started with the CF-100, an all-weather twin-jet fighter. They then astounded everyone by producing the first jet-powered airliner in North America, the C-102 *Jetliner*, powered by four Rolls-Royce

Derwent engines. The Canadian government ordered work to be stopped due to Korean War priorities, even though several U.S. airlines had expressed interest in buying it. Then came the CF-105 *Arrow*, a Mach 2 fighter. In 1958 it flew faster, further, and with more armament than any competitor, but the government cancelled the program! They said that the length and breadth of Canada could be adequately protected by the *Bomarc* missile, which turned out to be a dud, and that the aircraft was too expensive—another fallacy. The company, exasperated and disillusioned, closed down in 1959. It's an interesting story, and more details are given in a later chapter.

Post-War Airliners

In civil aviation, companies such as Boeing, Douglas, and Lockheed produced transports during World War II, albeit for the military, but easily adaptable to civil usage. From Boeing, there was the C-97 *Stratocruiser* developed by installing a new fuselage on a B-29 wing. From Douglas, there were the DC-3, DC-4, and the DC-6, and from Lockheed, there was the C-69 *Constellation*. No air transports have been built in the U.K. since 1939, so they had to buy U.S. aircraft or modify some bombers. The Avro *York*, for example.

An immediate post-war test of airlift capability occurred in 1948 when the Russians prevented land-based traffic from taking supplies into Berlin. The allies responded with a massive airlift, using all available transport types, and an airplane landing in Berlin every 3 minutes for 24 hours per day.

Taking advantage of their work in jet propulsion, de Havilland jumped ahead of the pack by flying the world's first jet airliner, the DH 106 *Comet*, in mid-1949. Within three years, it was flying passengers for British Overseas Airways Corporation (BOAC). However, in 1954 it ran into unforeseen problems when one exploded over Italy, and another exploded two weeks later. After a very extensive investigation of the wreckage, it was concluded that metal fatigue due to repeated flexing of the pressurized fuselage structure had caused cracks.

This was a relatively new phenomenon, and it caused some U.S. manufacturers to scrutinize their upcoming designs to ensure that they, too, would not encounter such problems. The *Comet* had been thoroughly ground tested at pressures far higher than those being actually used, but the <u>repeated</u> flexing of the fuselage had never been considered. After four years of beef-up and other changes, such as the landing gear, the *Comet* was back in operation. But it was too late to retain its dominance in the long-distance market. Boeing 707s and Douglas DC-8s were now on the scene. However, later *Comets* were still widely used by BOAC, Panair do Brazil, Canadian Pacific, RCAF, Japan Air Lines, Mexicana, BEA, and others.

In May 1967, a militarized version of the *Comet*, known as the *Nimrod*, made its first flight, and this was used for many years by the RAF for maritime reconnaissance. It was well-liked by the crews and was very effective in its missions. It had an endurance of 12 hours, was used to detect submarines, and was capable of launching torpedoes, bombs, depth charges, and mines.

As noted, the *Comet's* accidents gave the U.S. a chance to catch up. Five years after the *Comet's* first flight in May 1954, the first Boeing 707 was rolled out of the workshops. Its first flight was in July 1954, followed by the Douglas DC-8. They were somewhat faster and larger than the *Comet*, and the development of the 707 was helped by Boeing securing a contract for a USAF tanker version (KC-135) based on the 707 airframe. This contract provided Boeing with funds to enable them to continue the development of the airplane. In October 1958, the *Comet 4* had the honor of being the first jet airliner to provide trans-Atlantic service but was quickly followed by the 707 and DC-8.

On a smaller scale, Britain's Vickers *Viscount* became a bestseller, with 444 of them sold. It was powered by four Rolls-Royce *Dart* turboprop engines. I flew in one from Capital Airlines and was impressed by its low noise level and large upright-oval windows. It first flew in mid-1948 and predominated the small-airline market until the Douglas DC-9 came along in 1955.

Engine Development

At this point, it is useful to notice how engine development was the driving factor in these great leaps in aviation. During World War II, de Havilland produced the *Goblin,* used in the DH *Vampire* and the prototype Lockheed XP-80. The DH *Ghost* engine was similar to the *Goblin* but larger and more powerful. Bristol, in cooperation with Rolls-Royce, produced the *Olympus* and the *Pegasus.* Gradually, Rolls-Royce took over the jet engine field in the U.K. and Europe and is now one of the three primary companies in the world. The other companies are Pratt & Whitney (P&W), CFM and General Electric (GE) in the U.S. Since World War II, jet propulsion has made enormous progress. The DH *Goblin* had 3100 lb thrust and was succeeded by the DH *Ghost* with 5300 lb thrust. The P&W 119 developed 35,000 lb thrust for the F-22 fighter.

These improvements have been achieved by gradually changing from the original pure jet with a large-diameter centrifugal compressor to an axal compressor with multiple rows of very sophisticated compressor blades at the cost of greatly increased complexity. Another variant is the turbofan, which is now used extensively on commercial aircraft due to its superior (low) fuel consumption. With this variant, the fan is geared to rotate at a more efficient (slower) rate than the core engine. This results in tremendous thrust, but the frontal area is very high—many times the area of the core engine. Earlier turbofans were used on the Lockheed C-5 *Galaxy* and Boeing 747. Later models are used on airliners, such as the Boeing 787 and Airbus A380.

Despite the emphasis on jet propulsion, propellers have made great strides and are now grotesquely shaped with multiple blades. Used with turboprop engines, they are now widely used on transports such as the Lockheed C-130J and the Airbus A400.

The Cold War and Beyond

Potential, threatened, or ongoing war has been the primary mover in aviation development because defense departments provide fairly risk-free funds. During the "Cold War," which began with the Berlin Airlift and ended when the USSR broke up, the Lockheed U-2 Spy Plane was used to overfly the USSR and Cuba to search for missile sites. In the early 1960s, the Lockheed SR-71 flew faster than the Mach 3 and higher up to 85,000 feet than any other— typical of Kelly Johnson's—"Skunk Works"!

In the U.K., the Hawker *Hunter* was a very successful fighter. It entered service in 1954 and became widely used. Then came the V- bombers– the *Valiant, Vulcan, and Victor,* from Vickers, Avro, and Handley Page that entered service in 1955, 1956, and 1958 respectively . These V-bombers could carry nuclear weapons, and the *Vulcan* finally emerged as the backbone of Britain's bomber force. The English Electric *Canberra* was introduced in 1957, a twin-engine light bomber that was also built under license by Martin in the U.S. as the B-57. Altogether, some 1300 of these were produced, including 403 from Martin.

New jet bombers were also appearing in the U.S. The long-range, six- engine Boeing B-47 *Stratojet*, the Boeing B-52 *Stratofortress*, Convair B-58 *Hustler*, and Rockwell B-1 *Lancer*. Along the way was the North American XB-70 *Valkyrie* that was tested in the early '60s and cancelled in 1962 after only one prototype was built. The B-47, of which 2000 were produced, was the foundation for Strategic Air Command (SAC). It was the first swept-wing jet bomber and operated from 1948 until 1966. It had two unique characteristics: two inboard engines (GE J-35) in one pod and another engine outboard; Also, it had a bicycle-type landing gear with main gears beneath the forward and aft fuselage and outriggers near each wingtip.

The six-engine B-52 *Stratofortress* quickly became the mainstay of SAC and still is, although SAC has now become the Air Force

Global Strike Command (AFGSC). It has a 12,500-mile range, enabling it to carry up to 70,000 pounds of weapons at high subsonic speeds to almost anywhere in the world. The B-52's first flight was in April 1952 and entered service in February 1955. The Convair B-58 *Hustler* was a supersonic bomber but somehow didn't become popular with SAC. Its first flight was in 1956, and it was retired in January 1970. The Northrop-Grumman B-2 *Spirit* is the latest U.S. bomber. Its unique characteristic is that it has an all-wing design and has stealth properties. It first flew in July 1989, was in service by April 1997, and has been in operation in the Kosovo War, Iraq, and Afghanistan. It can carry up to eighty 500-pound JDAM bombs or sixteen 2400-pound B83 nuclear bombs. Without refueling, it has a range of more than 6,000 nautical miles.

Fighter aircraft have also made great strides since World War II. As noted above, the Hawker *Hunter* was one of the first new jet fighters in the U.K. Since then, the British aircraft industry has produced the *Harrier, Tornado,* and *Typhoon.* The *Harrier* was designed by Hawker as an outgrowth of the P. 1127 *Kestrel* and is undoubtedly one of the greatest innovations in recent years. Its *Pegasus* engine nozzles can rotate beyond 90 degrees allowing the thrust to be horizontal or vertical. It went into service with the RAF in 1969 and later, as the Sea Harrier, with the Royal Navy. It was used successfully in the Falklands War and was subsequently bought by the U.S. Marine Corps, then called the AV-8B, for ground attack, close air support, and fighter. As this is written, it is still used by the Royal Navy but will be replaced by the Lockheed F-35. The Panavia *Tornado,* produced as a joint venture between British Aerospace, MBB of Germany, and Alitalia, made its first flight in 1974. It has a variable-sweep wing and is used primarily as a fighter-bomber. Nine hundred ninety-two of them were built. Among the wars, it participated in the 1991 Gulf War, Bosnia/Kosovo, Iraq, and Afghanistan. The British Aerospace Eurofighter *Typhoon* is the latest multirole fighter from the U.K. and Europe, and like the *Tornado,* it was designed by a consortium of three companies based on previous BAE research: British Aerospace, Airbus, and Alena Aermacchi. Its first flight was in March 1974 and

33

entered service in August 2003. It has twin engines and can cruise at supersonic speeds. Tests show that it has remarkable agility, making it an excellent dogfighter. With such credentials, it has been purchased by the air forces of several countries, notably the RAF, Germany, and Italy, and was in action in the Libya conflict.

In the U.S., following the F-4 and F-16, Grumman delivered the F-14 *Tomcat* to the U.S. Navy in 1974 to replace the *Phantom*. Lockheed developed the A-12/YF-12A/SR-71 series and then the F-16 in 1974, the F-117 in 1981, the F-22 in 1997, and finally the F-35 in 2006. Each of these aircraft had some interesting, unusual, or significant features.

The A-12 incorporated early stealth concepts and first flew in 1965, and like all others in that series, had a top speed exceeding Mach 3.5. It was used for reconnaissance at very high altitudes. The YF-12 version was an interceptor, and the SR-71 *Blackbird* was the best-known version. All aircraft in this series had special skin surface treatment to minimize radar signature and utilized titanium for many structural areas to provide adequate strength at the high skin temperatures encountered at very high speeds. The SR-71 became operational in 1966, and 32 of them were built. They were used extensively over Vietnam, where their high speeds were enough to evade enemy aircraft or missiles. One interesting fact is they flew from New York to London in an hour and 56 minutes to attend the Farnborough Air Show!

The F-16 *Fighting Falcon* began its life at General Dynamics (GD) and then became a Lockheed design when GD sold its aircraft business to them. It made its first flight in 1974 and was introduced to service in August 1978. Since then, there have been 138 configurations. With sales to 11 countries due to its affordability and excellent capabilities—an internal Vulcan cannon plus 11 locations for other weapons—it has been one of the most popular modern fighters.

The F-117 *Nighthawk* came from Lockheed's Skunk Works and became operational in October 1988 as the world's first super-stealthy fighter. Not only did it have a radar-absorbing surface, but those surfaces were cleverly designed to be at various angles to each other. In addition, it had specially-designed engine intakes and exhausts to virtually eliminate its radar detection. With these odd shapes and engine treatments, it was not supersonic, and it really wasn't a fighter despite its "F" designation. Instead, it was used primarily to launch bombs of various types in the Gulf War and Iraq.

Next in line with Lockheed (now Lockheed Martin after its merger) was the F-22 *Raptor*. A stealthy Mach 2.2 air-superiority fighter that began operations in December 2005. It was designed to replace the F-117, F-15, and F-16 and is noted for its excellent maneuverability, enhanced by vectoring engine nozzles.

The latest Lockheed fighter is the F-35 *Lightning II* that first flew in December 2006. It is described as an all-weather stealth multirole fighter and has three versions—the "A" model for conventional flight, the "B" model for VSTOL, and the "C" model for carrier operations. Consequently, it is destined to be used by both the USAF and the Marine Corps, plus a number of foreign air forces. Notable foreign participants are the U.K., Italy, Canada, Norway, Denmark, the Netherlands, Israel, and Turkey. It can supercruise up to Mach 1.2 and is powered by a P&W 135 engine with a Rolls- Royce LiftSystem for vertical thrust, using vectoring nozzles. The F-35 armament is mostly carried in two internal weapons bays, but it has external hardpoints for other weapons. Its cost is relatively high due to its VSTOL capability and the ultra-sophisticated computerized systems needed for 21st-century combat.

Russian aircraft have not been mentioned in the discussions of bombers and fighters primarily because they haven't exhibited anything new or unusual beyond the U.K. and U.S. products. However, it is noteworthy that the Russian industry is constantly delivering aircraft of all types that are quite competitive with those from the western powers And this fact is responsible for the never-ending search for higher speeds, higher stealth, higher altitudes, improved armament, and, it seems, higher costs.

Cargo Planes

The Berlin Airlift was the first time that airplanes were used extensively to haul cargo, and that was primarily using the DC-3 or its military equivalent. These were not really designed for cargo operation. The first example of an actual cargo aircraft was the Lockheed C-130 *Hercules* which took off in August 1954 as the YC-130, and in 1955 the first production model flew. It was notable for its boxcar-type cargo compartment with straight-in loading and a ramp to enable trucks, bulldozers, and guns to be driven in. Also, it was capable of off-field performance, and there have been many versions ranging from "A" to "J." Thirteen of them were equipped with skis for Antarctic operation with the U.S. Navy. The "E" model was a significant improvement with a 30% increase in payload and a long enough range to permit transatlantic flights. The AC-130 Gunship proved itself in Vietnam; USAF Special Operations used the MC-130E and F for various covert missions. The C-130H was a major advance over the "E" model with an even longer range and updated avionics. The RAF used the C-130K, and they added a 180-inch stretch. Then came the "J" model with its increased performance and reliability. It made its first flight in June 1996. It was also stretched and has a new up-to-date cockpit with many automated functions, new and more powerful turbofan engines for improved performance, and a longer range. It can take off and land even shorter distances than its earlier models. The whole C-130 story is very interesting, discussed later in this book. Civil versions known as the L-100 have been produced with stretches up to 180 inches. Many were sold to large and small airlines to provide for the emerging air cargo market. It is currently the only commercial cargo plane in its class capable of carrying M-1 and M-2 containers and pallets up to seven 8 x 10-foot containers/pallets.

The C-141 *Starlifter* was the next step-up in cargo transport. First, as opposed to the C-130's turboprop engines, it had four turbojet engines. It retained the high-wing principle of the C-130 to enable it to sit low- to-the-ground for easy loading. Its long range allowed it to fly non-stop from California to Vietnam, thereby providing a tremendous asset to the U.S. military in that conflict.

The C-5 *Galaxy* was the result of a competition between Boeing and Lockheed. Boeing lost the competition but made great use of its studies to produce the ensuing Boeing 747. Lockheed went on to produce the C-5, and it flew for the first time in June 1968. It has proved to be a tremendous leap ahead in cargo transport. Powered by four GE TF-39 high bypass engines, it can take off at weights approaching 800,000 pounds and has many unusual features. It can be loaded from the front and back. It can kneel to permit straight-in loading from a conveyor, or vehicles/guns can be driven straight in up the ramp. It is large enough to accommodate up to six large buses or two of the U.S. Army's main battle tanks, and it has an upper deck large enough to hold 75 fully-equipped troops in addition to the cargo below. Furthermore, its unique 28-wheel landing gear can operate from unprepared surfaces. It has also air-dropped a record of 80 tons of material. In late 1982, Lockheed obtained a contract for an upgraded version, the C-5B. That was followed by the "C" and then the "M" models. The latter, which became operational in 2014, has a 22% increase in engine thrust that has reduced takeoff distance by 30%, and payload has gone up to 176,610 pounds.

The Boeing C-17 *Globemaster III* flew in September 1991 after a competition for an Advanced Medium STOL Transport (AMST) to, supposedly, replace the C-130. It is powered by four P&W F117 turbofans and has a payload of 170,900 pounds, such as one M1 main battle tank. It can carry 120 paratroops with their equipment and operate from relatively short runways (3,500 feet) at a reduced weight. Operating extensively from Iraq, Afghanistan, and other hot spots, it is well-liked by the crews and flown by the USAF, Royal Air Force, and air forces of Canada, Australia, and several other countries.

The latest cargo plane is the Airbus A400M. It originated from studies performed by a group from various companies—Aerospatiale, British Aerospace, Messerschmitt-Bolkow-Blohm, and Lockheed. The latter company eventually left the group to study other airlift alternatives. The project was known as Future International Military Airlifter (FIMA) and was intended to be a C- 130 replacement. After several years of study, the design was finally completed, and the A400M took off for the first time in December 2009, with the initial delivery to the French Air Force in August 2013. As expected, it is an intermediate between the C-130 and C-17, has four Europrop TP400-D6 engines, a payload of 81,600 pounds, a gross weight of 310,652 pounds, and cruises at 485 mph (Mach 0.68). When carrying 100 tonnes payload, it has a takeoff distance of 3,215 feet.

Finally, there is the Bell Boeing V-22 *Osprey* tilt-rotor vehicle. It is not a conventional airlifter, and it is considerably smaller than the others, but it DOES carry cargo and must be recognized because it is so unusual. First, it can take off and land vertically using tilting engines on the wing tips. These two Rolls-Royce AE 1107C engines drive huge proprotors. It began operations with the U.S. Marine Corps in 2007 and with the USAF two years later. Since then, it has proved itself valuable in Iraq, Afghanistan, Sudan, and Libya, deploying and extracting troops and/ or cargo from areas that are not accessible to other types of airlifts and, in some cases, ground transportation. It can haul up to 32 troops or 20,000 pounds of cargo and has a top speed of 277 mph.

The latest airliners

As noted earlier, the Boeing 747 was developed from Boeing's C-5 cargo plane proposal. Pan American was a strong advocate of the 747, but some airlines thought it was too big. This precipitated a spiral of designs from other manufacturers. TWA, United Airlines, and Eastern Airlines supported Douglas's DC-10 and Lockheed's L-1011 wide-bodies. Both are quite similar and have three engines. The airlines liked these because they were the "right" size and had the safety of having three engines rather than two for over-water flights. In the U.K., de Havilland also produced a tri-jet, the DH 121 *Trident*, but this was overshadowed by the Boeing 727 that appeared shortly afterward. The French industry, namely Airbus, sprang to life in 1970, and its A300 generated interest and orders from Eastern Airlines. The French and British governments decided to fund a supersonic airliner, the *Concorde*, entering service in March 1976. Back at Boeing, revitalized their production line with the 737, 757, 767, 777, and 787. So within 40 years, there had been a tremendous evolution in commercial air travel.

The Boeing 747 was the first in line, entering service in January 1970. Its enormous interior and upper deck can carry 416 to 660 passengers depending upon the seating configuration, providing them with hitherto unknown luxury for long-distance flights. So, it is well-liked by both the passengers and the airlines, making it a very successful program.

The DC-10, which started service in August 1971, had a maximum passenger capacity of 380. Still, it had a series of accidents that upset passengers, resulting in declining orders, expensive corrective actions, and program termination. The L-1011 went into service a year after the DC-10 and was particularly well-liked by the crew. Although it was a popular airliner and carried about the same number of passengers as the DC-10, Lockheed ran into financial difficulties and decided to exit the airliner field.

Airbus has thrived, gradually progressing in wide-body airliners. From the A300, which could accommodate 266 passengers and started operations in 1974, to the A-380. The latter is the world's largest commercial aircraft and can carry between 400 and 800 passengers depending upon the configuration. They are truly competitive with the Boeing planes, and the news media regularly compares orders obtained by these two companies at various international air shows. Overall, it is a manifestation of how customers benefit from competition.

The *Concorde* was probably ahead of its time. Design studies began at Britain's Royal Aircraft Establishment in the early 1950s and later included Hawker Siddeley and Bristol. Then in the late 1960s, three French companies started examining the possibilities of a Supersonic Transport (SST). The following year the British and French companies worked together, and in 1962 a treaty was signed by the two governments to jointly develop the aircraft. It made its first flight in March 1969 and made its first transatlantic flight on September 4, 1971. The *Concorde*, operated by Air France and British Airways, could accommodate 92 to 120 passengers depending on the seat density chosen. But compared to current transports, the interior seems to be somewhat cramped. However, the speed-comfort factor should be considered, that is if you're going to reach your destination in half the time, you can be quite satisfied with less space! The ocean liner is the other extreme . It has super comfort but takes several days to cross the Atlantic.

Boeing's latest airliner is the 787 *Dreamliner* which made its first flight in December 2009. It is noted for its particularly attractive and comfortable interior and its excellent fuel efficiency. Passenger capacity ranges from 210 to 325, with 234 being the typical load in three classes, putting it in the popular mid-size group, and it has a long range. Composite materials are used for most of their structure. This, plus efficient GEnx or Rolls Royce Trent 1000 engines and refined aerodynamics, provides fuel efficiency 20% better than the Boeing 767. Several problems occurred during its early life, primarily with the lithium-ion batteries. But, these have been overcome, and it is now enjoying successful operation with several airlines beginning with All Nippon Airways in October 2011.

Biz Jets

In late 1960, large corporations (and some government entities) began to realize that they could profit by using their own small jet airplanes to visit clients or transport VIPs, to attract customers by appearing with their jets at air shows, and merely travelling fast and comfortable between their various branches and subcontractors. While en-route, they could also attend to business, have small meetings in the cabin, and be able to meet colleagues within a short time and return home that same day. Grumman designed the Gulfstream, and it made its first flight in 1958. Later, it was updated and redesigned several times by Gulfstream Aviation such that it is now being widely used because of its transoceanic range and relatively large cabin. The latest version of the Bombardier Global 5000 is similar to the Gulfstream and provides stiff competition. About the same time as the original Gulfstream was produced, Lockheed produced the twin-engine JetStar, later modified to have four jets. It was used by the USAF and commercial customers. North American flew their twin-jet Sabreliner in September 1959. The Learjet emerged as a very popular twin jet in 1963 and is still in production. In the U.K., de Havilland (later Hawker Siddeley) designed their DH 125 twin-jet, a well-liked airplane that flew in 1962. In France, Dassault produced their first Falcon in 1977, an airplane that has sold widely in both two- and three-engined versions.

That concludes the overview of aviation from my perspective, which began with two-seat open-cockpit biplanes. Since then, some 75 years ago, we've seen airplanes that can now zip across the Atlantic in a few hours, carrying hundreds of passengers in a comfortable environment at fares that are acceptable for almost everyone. In the military sphere, we now see some functions being taken over by unmanned vehicles and missiles being used, in many cases, to replace fighters. It is indeed an exciting time!

The following chapters describe some of the exciting, unusual, and sometimes dramatic events in aviation history.

Short Empire Flying Boat

Boeing 314 Flying Boat

Boeing B-17

Lockheed P-38 Lightning & F-22 Raptor

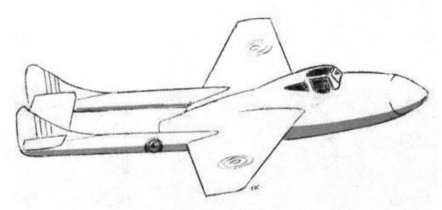

de Havilland Vampire

Boeing B-29

North American F-86 Sabre

Lockheed F-80 Shooting Star

Avro Canada CF-100

Avro Canada Jetliner

Boeing 707

Boeing 737

Boeing 747

Airbus A380

Airbus A340

Airbus A300

Northrop Grumman B-2

LockheedMartin SR-71

McDonnell Douglas F-15

LockheedMartin F-35

LockheedMartin F-22

LockhedMartin F-16

Boeing B-52

Hawker Harrier

LockheedMartin C-130

LockheedMartin C-141

LockheedMartin C-5

Boeing C-17

CHAPTER 3
TRANSATLANTIC FLIGHT

Most readers know about Charles Lindbergh flying across the Atlantic, and some may know the names, Alcock and Brown. While they were "firsts" in some categories, they were not the first to fly across the Atlantic. Instead, two other names, Read and Hinton, get top billing, and chances are that few readers have ever heard of them or their airplane, a Curtiss NC-4. It was a flying boat, about 68 feet long, had four 400 hp Liberty engines, weighed 28,500 pounds, and had a top speed of 90 mph. Crew accommodations were minimal. The pilot sat in the open behind an automobile-type windshield, and the navigator stood in an exposed hole in the front of the fuselage.

After the end of World War I, the Royal Navy purchased some flying boats from Curtiss in the U.S. and named them *Felixtowes*. They modified them to obtain a longer range and have better engines and hulls. Curtiss liked these changes and incorporated them into their design which ultimately became the NC-4.

The U.S. Navy bought the NC-1 through NC-4 flying boats and decided to attempt a transatlantic flight with them, particularly, the NC-4. Navigated by Lt. Commander Albert Cushing Read, and piloted by Lt. Walter Hinton, Lt. Stone and Lt. James L. Breese, they took off from Naval Air Station Rockaway on May 8, 1919, accompanied by two other NCs. After stopping at Naval Air Station

Chatham (Massachusetts) and Halifax, Nova Scotia, they arrived at Trepassey, Newfoundland, the starting point for their transoceanic flight. Up to that point, eight U.S. Navy ships were positioned along the east coast to help with navigation.

By May 16 they were ready for their long flight, and 22 U.S. Navy ships had been stationed along the route to the Azores for navigation and rescue purposes. In addition to pilot and navigator, the crew members were Chief Eugene Rhodes, flight engineer and Herbert C. Rodd, radio operator. The two flying boats accompanied them but had to land in the Atlantic due to poor visibility. One sank but with no casualty, and the other was damaged when it too landed in the ocean and then taxied to the Azores.

Bad weather delayed their takeoff from Newfoundland, and rough seas cut their speed so that takeoff was difficult. Once they were on their way, they found that navigation was difficult due to fog, clouds, and rain. First, in the heavy fog, there was the possibility that the three airplanes may hit each other, so they decided to abandon any ideas of flying in formation. By climbing slowly above the fog, they could navigate by the stars. The commander of one of the airplanes, Lt. Cmdr. Bellinger was aware that he was getting low on fuel and was close to the Azores, so he decided to touch down in the ocean to check his position. However, when he penetrated the mist, he discovered that the waves were 20 feet high. But it was too late; he landed. They had ham sandwiches and coffee and contemplated their fate, eventually sending out an SOS. A Greek freighter, the *Ionia*, came by and picked them up. They were transferred a little later to the U.S. destroyer *Gridley*. After trying unsuccessfully to tow their airplane, they sank it to avoid shipping hazards. They then discovered that they were only 100 miles from the Azores.

Under Commander. John Towers, had a similar experience as they battled rain and wind, with few glimpses of the sky to check navigation. Towers was also told about his remaining fuel so he decided to land and check his position. Like Bellinger, he found that the waves were at least 20 feet high too late. The flying boat hull was damaged during the landing. Towers estimated that they were close to the Azores Island of Corvo, so they decided to taxi to it. The pounding waves broke off the pontoons from each wing meaning that the wing tips were liable to touch the ocean, thereby breaking off the wings. To counteract the imbalance, crew members climbed onto the wings and darted back and forth to keep them level. Eventually, they sighted land, and shortly after that, the U.S. destroyer *Harding* appeared and escorted them to Ponta Delgada in the Azores, where they were given a 21-gun salute!

Lt. Commander Read's NC-4 also encountered navigation difficulty. He stood in the bow some five feet in front of the pilots and waved his arms to show them whether to go up, down or change their lateral attitude. In such poor visibility, this was their only way to communicate! When his calculations showed that they were almost at their destination, they descended through the clouds and saw the land. Hence, they proceeded for another six minutes to the harbor at Horta in the Azores, where they pulled up alongside the USS *Columbia* and were piped aboard!

It had taken them 15 hours 18 minutes to reach the Azores, and after three days, they left again for Lisbon, Portugal. Mechanical trouble forced them to land at Ponta Delgada, and after they'd made repairs, they arrived on May 27. Once again, the Navy had ships en route to help with navigation, and eventually, they landed in Lisbon. Their first question there was: were they the first to make that crossing, or had one of the other teams back in Newfoundland beaten them? These other teams were flying land planes and had assembled at St. John's about 50 miles from where they had started at Trepassey Bay. The Captain of a U.S. destroyer shouted to them that they were the first. It transpired that their most serious competition, Harry Hawker with his navigator MacKenzie Grieve from the U.K., had crashed after flying 1100 miles across the Atlantic and were picked up by a passing steamer. Another team, Alcock and Brown were still waiting to take off.

So THIS WAS THE FIRST AIRCAFT TO FLY ACROSS THE ATLANTIC.

On May 31, 23 days after leaving Newfoundland, they flew to Plymouth, England. As expected, there was considerable excitement, and this continued later in London and Paris. After celebrating, the airplane was taken back to the U.S. by ship.

At this point we recognize a 1913 statement in the London *Daily Mail*. It offered a prize of 10,000 pounds which was a lot of money back then to *"the aviator who shall first cross the Atlantic in an aeroplane in flight from any point in the United States of America, Canada or Newfoundland and any point in Great Britain or Ireland in 72 continuous hours."* The U.S. Navy decided that the NC-4 couldn't do it. Still, several intrepid teams decided to "have a go." The only successful one was the Vickers team with Alcock and Brown flying a modified Vickers Vimy World War I bomber.

Captain John Alcock was born in 1892 in Manchester, England, and obtained his pilot's license in 1912. He joined the Royal Flying Corps, the forerunner of the RAF, in World War I as a bomber pilot. Lieutenant Arthur Whitten Brown was born in 1896 in Glasgow, Scotland, to American parents. They moved to Manchester, and after training as an engineer, he also joined the Royal Flying Corps as a fighter pilot. He later developed skills in navigation.

After the war, Vickers in Weybridge, England reviewed the *Daily Mail* offer, suspended during the war but was now reinstated. They felt that their Vimy bomber could win that prize with modifications to increase its range. It had two 360-hp Rolls Royce Eagle engines, was about 44 feet long and had a wing span of 68 feet. At that time, Alcock agreed to be the pilot. Brown was unemployed and asked Vickers for a job, and after noting his capabilities, they offered him the position of navigator on this project, to which he agreed.

So it was that, on June 14, 1919, at about 1:45 pm, Alcock and Brown took off from St. John's, Newfoundland, and flew across the Atlantic to Clifden, Connemara, County Galway, Ireland. They flew below the clouds and generally at low altitudes most of the time. Flying time from Newfoundland to flying over land again was about 16 hours, and by excellent navigation, they were close to where they expected to be. Their chosen landing site appeared to be a nice grassy field, but right after they touched down, they discovered it was an Irish bog causing the airplane to pitch over onto its nose. It was damaged, but they escaped without injury. They had flown 1890 miles in 15 hours 57 minutes at an average speed of 115 mph.

THIS WAS THE FIRST NON-STOP FLIGHT ACROSS THE ATLANTIC OCEAN

After the accolades had subsided, they received their *Daily Mail* prize from Winston Churchill, who was then Secretary of State for Air, plus some others, and flew to Manchester to be honored by the Lord Mayor and City Council. They were knighted by King George V a week later.

On December 18, 1919, Alcock died in France on due to an accident while flying a Vickers Viking flying boat to the Paris Aircraft Exhibition. Brown got married, and in World War II was a Lt. Colonel in the Home Guard before rejoining the RAF to help train navigators. His only son, Arthur ("Buster"), was killed while serving in the RAF when his *Mosquito* crashed in the Netherlands in 1944. Sir Arthur died in his sleep in Buckinghamshire on October 4, 1948.

The next transatlantic pioneer of consequence was **Charles Lindbergh.**

Charles was known sometimes as Lucky Lindy, but the most suitable name was "Slim," for he was a tall, slim guy, shy and usually soft-spoken. Born in Detroit on February 4, 1902, he grew up on a farm in Little Falls, Minnesota, and his father was a U.S. Congressman for Minnesota from 1907 to 1917. Charles exhibited natural mechanical skills, and at age 18, he began studying engineering at the University of Wisconsin. Two years later, he left and became a barnstormer performing dare-devil stunts at fairs. He joined the U.S. Army in 1924, and the following year he graduated from its flight training school. Robertson Aircraft Corporation then hired him to fly airmail between Chicago and St. Louis, thus giving him some valuable experience.

Back in 1919, a New York hotel owner, Raymond Orteig, had offered a $25,000 prize to the first person who flew nonstop from New York to Paris, and needless to say, several aviators became interested. Some died trying to do it. Lindbergh felt sure that he could win the prize if he had the right airplane and the money to build it. He succeeded in getting some St. Louis businessmen to provide him with $15,000, and he prepared the rough details of the airplane he needed. He approached several companies such as Fokker, Wright and Bellanca to ask them to design and build his airplane, but with that small amount of money, they declined. He then contacted a new company, Ryan Airlines, Inc. of San Diego, California.

Lindbergh wanted a monoplane with a single air-cooled engine that could fly 4,000 miles, and he wanted it in three months. Ryan said they could do it, so in February 1927 he visited the plant, a decrepit-looking place on the waterfront. He talked to their top guys, including Don Hall, their engineer. Total staff at Ryan numbered 35, including everyone from the owner to the secretaries, so, understandably, Charles was somewhat apprehensive about their capabilities.

Hall got to work on the design and soon had it figured out. He presented it to B. F. Mahoney, Ryan's owner, who asked if it could be built in 60 days, 30 days less than specified. Hall thought it could, so Mahoney agreed to take on the job. The timing was critical, though, since the other aviators made good headway with their airplanes to win the Orteig Prize. They included Major General Patrick of the U.S. Army Air Corps, Commander Byrd of the U.S. Navy, and three Frenchmen, Rene Fonck, Charles Nungesser, and Francois Coli.

Work progressed quickly at Ryan, but the workers were upset by Lindbergh's constant presence, meticulously checking everything and how things were done. For example, on such one-off projects as this, it would be normal to cut off and bend hydraulic tubing to fit, but Lindbergh told them to limit tube lengths to 18 inches and to connect them with rubber hoses. This was to avoid oil line breakage or leaks due to vibration. When asked why he wanted such perfection, Lindbergh replied: "Because I'm a damn poor swimmer, that's why!"

One interesting feature that was added to the airplane was a periscope. An elderly fellow named Randolph suggested it. He'd been a submarine captain and had noticed how the pilot's forward vision was restricted by a gas tank and engine cowling. Lindy thought this was a great idea and it significantly benefited the operation.

In addition to manufacturing perfection, Lindbergh was obsessed with minimizing weight. Every ounce added would decrease his range and increase takeoff distance—and the latter proved to be vital later on. He even designed and made his own lightweight flying boots and removed any parts from his charts that were not needed.

Eventually, the airplane, which he'd named the *Spirit of St. Louis*, was completed. Ryan employees had worked long and hard hours, but they finished it on time, and despite their early frustrations with him, they were now solidly and enthusiastically behind his effort. On April 26, 1927, the *Spirit of St. Louis* made its first flight, and all went well. After more tests, Lindy flew it to St. Louis in the record time of 14 hours and 25 minutes. From there, he proceeded to Roosevelt Field near New York City.

At 7:52 am EST on May 20, 1927, Lindy finally took off from Roosevelt Field. This was more hazardous than he anticipated because he had more fuel onboard than he planned, exceeding i153 pounds more on the weight. The airfield was soft and muddy from heavy rains, so takeoff was impeded, and in addition, he was taking off downwind. The latter was caused by taxiing to the other end of the field, which may cause the engine to overheat, and taking off into wind would have taken him over hangars and houses in which a crash would have been fatal. When he did take off, he barely cleared the telephone lines at the end of the field before heading out over the ocean.

Twenty-eight hours after leaving New York, he crossed over Dingle Bay, Ireland. He flew over southwest England and Cherbourg, France after 32 hours. He landed at Le Bourget, Paris, at 10:21 pm on May 21 having made the 3600-mile trip in 33 ½ hours. Thousands of people were there to greet him, and there were the inevitable ceremonies and parades. President Calvin Coolidge presented him with the Congressional Medal of Honor and the Distinguished Flying Cross for his heroic flight. Afterwards, he took his airplane on a Central and South America tour and to most parts of the U.S.A.

AND THIS WAS THE FIRST SOLO FLIGHT FROM NEW YORK TO PARIS.

The next person in the story of Transatlantic Flight is **Amelia Earhart.** A month before her 31st birthday, she flew across the Atlantic in a Fokker F7 *Friendship* airplane, piloted most of the time by copilots Bill Stultz and Louis Gordon. They flew from Newfoundland to South Wales in 20 hours and 40 minutes on June 18, 1928, about a year after Lindy made his flight. Amelia became known as "Lady Lindy".

With the confidence from her first flight across the ocean, she decided to do it again. By herself this time! So, on May 20, 1932, she climbed aboard her Lockheed *Vega* in Newfoundland and crossed the Atlantic again. She had some difficulties en route. Visibility was poor, and when her altimeter stopped working, she had a problem knowing how far she was above the hard-to-see ocean. She also encountered some icing and had an oil leak, which caused her to abandon her plans to land at Southampton, and instead land in Ireland. Nevertheless, she made it and became the **first person to fly twice across the Atlantic.**

69

Transatlantic flight operations moved quickly after this. British Imperial Airways and Pan American Airways competed. Pan Am started weekly passenger service on June 24, 1939, leaving New York on Saturdays at 7:30 am, arriving in Southampton at 1 pm Sunday. Imperial Airways began its service in August of that year, with both airlines using flying boats. The British used Short Brothers Empire boats, the S 26 G-class, and Pan Am used the Boeing 314. However, World War II intervened, and most passenger flights were suspended from then until late 1945. Transatlantic flights were, however, continued by military aircraft, primarily those operated as ferry flights, in which U.S. aircraft were being constantly flown to the U.K., often using airfields in Newfoundland, Labrador, Greenland and Iceland along the way.

Curtiss NC-4

CHAPTER 4
POST & EARHART

With the Atlantic challenge taken care of, the Briitish turned their attention to their Empire routes, South Africa and Australia. The Germans were busy building up their air force for their future takeover of Europe, and the French didn't seem to be interested in further explorations by air. However, in the United States, there were two people wanted to fly around the world—Wiley Post and Amelia Earhart.

Wiley Post was born in Grand Saline, Texas, on November 22, 1898. His family moved to Oklahoma and bought a farm there, but Wiley was not content with that kind of life . He left school after the sixth grade and wanted to be a pilot in World War I; however, the war ended before he completed his training. So, as a restless 20-year-old, he worked in the oil fields, and at one point, he was arrested for car-jacking which cost him 14 months in jail. As a result of an oilfield accident in 1926, he lost his left eye and used the insurance money to buy an airplane! He became the pilot for oilman F.C. Hall who had purchased an open- cockpit Trans-Air biplane, and later a Lockheed *Vega* that he named *Winnie Mae* after his daughter. Wiley flew that airplane in the National Air Race Derby in August 1930 from Los Angeles to Chicago and won it, making him a celebrity. The *Vega* was an outstanding aircraft designed by John Northrop and Gerard Vultee; it was fast and rugged. It had a monocoque structure with plywood skins wrapped around wooden frames, and measured about 28 feet long, with a wing area of 275 square feet. Its gross weight was 4,033 pounds, and its maximum speed was 170 mph. It cruised at 140 mph and had a range of 725 miles. Between 1927 and 1934, 132 of them were built, and they were widely used by aviators at that time, including Amelia Earhart.

In 1931 Post flew his *Vega* around the world with Harold Gatty (an Australian) as navigator. On June 23 of that year, at 4:56 am, they took off from Roosevelt Field, NY, flew to Harbor Grace, Newfoundland, and crossed the Atlantic to Sealand Aerodrome, Chester (U.K.). Then on to Hanover and Berlin, Germany. They flew over Russia by way of Omsk, Novosibirsk, Irkutsk, Blagoveschensk, and Khabarovsk. Next stop: Nome, Alaska, and over the mountains to Edmonton, Alberta, Canada, and finally to Cleveland and New York, a total distance of 15,474 miles.

The only significant problem they had was caused by him trying to take off from a soft beach in Alaska. The propeller dug into the sand and was damaged. Using a hammer and wrench, he straightened the prop sufficiently to fly to Fairbanks where he could buy a new one.

After the flight, they received appropriate accolades, including lunch at the White House and a ticker-tape parade in New York City on July 7. Up to that point, his airplane was still owned by F. C. Hall, so Wiley bought it from him and made some changes, such adding a Sperry autopilot and a radio direction-finder.

In 1933, he made a second flight around the world—this time on his own. He flew out of Floyd Bennett Field, New York, on July 5, and after encountering bad weather over the Atlantic, he landed in Berlin. From there, he flew to Moscow and more or less copied his previous route over Russia. He landed in Flat, Alaska, a small mining community with a 700-foot landing strip, and damaged his prop again and the right-hand landing gear. Some local people could repair the landing gear, but he had to wait there until the propeller was repaired in Fairbanks. On July 22, he was on his way again to Edmonton, and finally a 2,000-mile, non-stop flight to New York, where he landed just before midnight.

His next project was high altitude flight where he joined forces with B.F. Goodrich to develop a pressure suit that would allow him to operate at altitudes up to 40,000 feet. They made three suits, one of which was successful. He would wear long underwear; on top of that was a suit with an inner black rubber air bladder and an outer layer made from rubberized parachute fabric. He also wore an aluminum/plastic diver's helmet. Further work on this project ended when he was killed in a crash on August 15, 1935.

Wiley had become good friends with a fellow Oklahoman and famous humorist Will Rogers who asked Wiley to fly him to Alaska to obtain material for his newspaper columns. At that time Wiley worked with Lockheed to provide him with a hybrid aircraft. This would combine parts from Lockheed's *Orion* and *Explorer* airplanes, and floats would be used instead of a landing gear. Lockheed told him that his proposal was unsafe and had balance problems. Still, Wiley decided to go ahead anyway and had the hybrid built by Pacific Airmotive Ltd. With this airplane, Wiley and Rogers set off on their Alaska trip. After leaving Point Barrow, they became unsure of their location and landed in a lagoon temporarily. While trying to take off again from that lagoon, they encountered an engine problem and stalled. The airplane pitched sharply nose-down into the lagoon, taking off the right wing, killing both men. And as noted earlier, this was on August 15, 1935.

As discussed in the previous chapter, **Amelia Earhart** began her adventures in that general time frame. This is the rest of the story. She and Wiley Post were both born in 1898, and she flew across the Atlantic three years before Wiley. They both attempted to fly around the world, and while Wiley succeeded in both 1931 and 1933, Amelia tried in 1937 and died in the process.

She was born on July 24, 1898, in Atchison, Kansas. Her dad was a railroad attorney who had to locate in different cities, so she went to school in Kansas City, Des Moines, and Chicago, where she graduated from high school. She attended Ogantz School for Girls in Chicago and visited her sister, Muriel, in Toronto, Canada, three years later. After seeing a large number of injured World War I soldiers walking the street there, she quit the Ogantz school and went to work as a Red Cross nurse's aide at Toronto's Spadina Military Hospital. She saw her first airplane right after the end of that war and spent time watching them take off and land and talking to pilots and mechanics. Her sister moved to Massachusetts to attend Smith College, so Amelia joined her there for a while and then went to study pre-med at Columbia University and Barnard College. She tired of that and moved to California, where her parents were then living, and it was there that she went on her first flight with Frank Hawks as the pilot. Hawks was an aviation pioneer and barnstormer in the 1920s and, later, an aviation consultant. This event changed her life. She she became so enthusiastic that she was determined to become an aviator.

She took a job at the Los Angeles Telephone Company to earn enough money to pay for flying lessons and obtain her pilot's license. Soon afterward, in a small Kinner Canary airplane, she broke the woman's altitude record (14,000 feet). She returned to academic life for a while, going to Harvard in 1925 and doing some teaching. Still, she managed to squeeze in some flying with a local chapter of the National Aeronautic Association. Then in 1928, Captain H. H. Bailey called and asked her she'd be interested in making a transatlantic flight. Apparently, a certain lady wanted to promote women's prestige in aviation and had planned to accompany a pilot and navigator flying across the ocean but had to back out for personal reasons. That lady wanted someone to take her place. Amelia was told to contact George Palmer Putnam in New York if interested. She was, and she did! She discovered that "the lady" was the Honorable Ms. Frederick Guest, a Pittsburgh native living in London. And so it was, Amelia accompanied pilot Wilmer Stutz and mechanic

Lou Gordon on the Fokker *Friendship* on the transatlantic flight discussed previously.

To recap that memorable flight, they flew out of Newfoundland on June 28, 1928, and arrived in the U.K. 20 hours and 40 minutes later. On May 20, 1932, she did it again, only this time Amelia was alone, and like Wiley Post she flew a Lockheed *Vega*. Due to some problems, she had to land in Ireland. An amusing conversation is supposed to have taken place at the end of the landing with the Fokker Fellowship when a local policeman rowed across to where Amelia was waving from the airplane where it was moored to a buoy. He said: "Do ye be wanting something?" The crew replied, "We've come from America," to which the policeman said, "Have ye now? Well, we wish ye welcome, I'm sure." In Paris, she was called "The Lady Lindbergh," and in London, she was the toast of the town.

In 1931, between her first and second Atlanta flights, Amelia married the George Palmer Putnam, an explorer, and publisher when their romance blossomed after he'd published her first book. It remained a good solid marriage. He supported her flying endeavors and helped her plan her second transatlantic flight.

Her next major trip was from Honolulu to Oakland, California, in 1935, with a distance of 2,408 miles. She had her Lockheed *Vega* shipped to Honolulu. On January 11, in the late afternoon, she flew out of there. Radio beacons helped keep her on course. After 18 hours and 6 minutes she landed at Oakland—**the first to fly from Hawaii to California.** On May 9 of that same year, she did "another first." She flew in a non-stop flight from Mexico City to New York, taking her over the mountains and across the Gulf of Mexico to the New Orleans area and then by way of Atlanta and Washington to Newark, a distance of 2,125 miles.

Amelia then joined the staff of Perdue University as Counsellor in Careers for Women, and the University presented her with a 10-passenger flying laboratory version of the Lockheed *Electra*—a new all-metal, twin- engine airplane. With the *Electra*, she thought it would be suitable for a flight around the world and that she could use it to increase knowledge of aviation medicine during that flight.

She made detailed preparations and took off from Oakland on March 17, 1937. She had Captain Harry Manning as navigator and Fred Noonan as his assistant. They arrived in Honolulu 15 hours and 47 minutes later—a record. Bad weather held them until the 20th, but then they were off again, headed for Howland Island in mid-Pacific, but a crosswind that caused the right wing to dip, and Amelia tried to correct it by pulling back on the left throttle rather than using the rudder. The airplane swerved to the left in a ground loop, collapsing the landing gear and breaking the propellers. Captain Manning had to return to the U.S. because his leave was expired, but Noonan decided to stay with Amelia. After repairs to the airplane by Lockheed in Burbank, they tried once again to make their round-the-world flight, only this time they changed the route, going eastward instead of taking the Pacific route.

On June 1, 1937, they left and flew first across the Caribbean to San Juan, Puerto Rico. After refueling, they went to Caprito, Venezuela, and over the rain forests to Paramaribo, where they spent the night at a hotel. The next day they flew to Fortaleza, Brazil, and Natal, after which they crossed the South Atlantic. It seems that they disagreed somewhat on navigation because she ignored Noonan's advice that she was off-course and consequently made landfall 110 miles north of their intended target of Dakar and finally landed at St. Louis, 163 miles north of Dakar. However, all went well, and they flew down to Dakar the next morning. They spent two days there and then flew across Africa to Khartoum. They went over some high mountains to Massawa on the Red Sea. Then down the coast to Assab and Karachi, where they had the engines checked while they relaxed for a while. She phoned her husband, who was back in America, and told him that they planned to be back there in time for the July 4th celebrations.

They took off for Calcutta on June 17. They went to Akyab to refuel, where they encountered a severe monsoon. On their second attempt, they reached Rangoon, where they were delayed again by monsoon weather. From there, they went to Bangkok and Singapore, after which they flew to Java, where they spent three days before flying over the Timor Sea to Darwin, Australia. They had the airplane thoroughly checked at this point because the most hazardous part of the trip lay ahead of them—crossing the Pacific.

In the morning of July 2, they left Darwin and refueled at Lae, New Guinea, but had no navigational aids from this point onward. They were too far from Australia to pick up radio beacons and did not have appropriate radio equipment to communicate with ship-to-ship transmissions. Therefore, negating any possibility of them receiving bearings from U.S. Navy and Coast Guard ships, as they tried to find Howland Island, a mere 2-mile-long strip of land in mid-Pacific. Navigator Noonan couldn't even use his sextant to determine their position due to overcast skies. The net result was that they didn't arrive at Howland.

The U.S. Coast Guard cutter *Itasca* and Navy tug *Ontario* patrolled the Howland Island area to help with navigation, but the absence of 500 kilocycle radio on the airplane prevented them from talking to Earhart/Noonan. On July 2, Coast Guard HQ in Washington received the following message from the *Itasca:*

"Earhart unreported at Howland at 7 pm (EST). Believe down shortly after 5 pm. Am searching possible area and will continue."

This precipitated a whole fleet of ships to be deployed by the U.S. Navy, including the battleship *Colorado* and the aircraft carrier *Lexington* with its 72 airplanes. A further message from the *Itasca* suggested that the flyers may have missed the island altogether for various reasons, and that they had flown northwest at about 3:20 pm EST (8 am local time). The sea was calm, visibility was good, but the dazzling rising sun could have blinded them. They had received a message from Earhart that she was about 100 miles from Howland and only had about a half-hour of fuel left, with no land in sight. Experts said that the Electra *Electra* could have floated for an extended period in calm seas with the landing gear retracted and almost empty tanks.

Onboard the aircraft, they had an assortment of safety measures, including a collapsible dinghy, an orange kite, and flares.

Messages from Amelia were reported by radio operators including a Pan American Airways operator and a British cruiser, but all of this came to nought, and as of today their disappearance remains a complete mystery. Several theories and a few facts have come forth since then. A British Colonial Service (BCS) officer took a poorly defined photograph in 1937 that showed what appeared to be an airplane landing gear protruding from some wreckage. Gerard Gallagher, another BCS officer, found some human bones on the island of Nikumaroro in the Pacific Island republic of Kiribati. Ten expeditions from the International Group for Historic Aircraft Recovery (TIGHAR) have found some other items on that island, such as a piece of aluminum, the heel of a woman's shoe, and a sextant box, leading them to conclude that this was, indeed, the island where they "landed." It is located about 300 miles southeast of Howland Island, and in 2013 some SONAR observations showed evidence of what could be airplane parts off the coast of that uninhabited island. So, at this point, it appears very likely that Amelia Earhart and Fred Noonan became castaways on the island of Nikumaroru and died there.

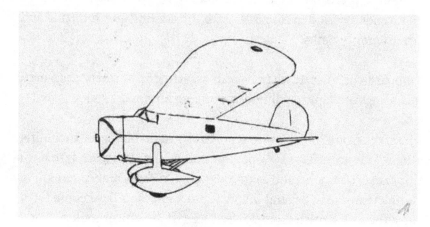

Lockheed Vega

Lindbergh's Spirit of St.Louis

Lockheed Electra

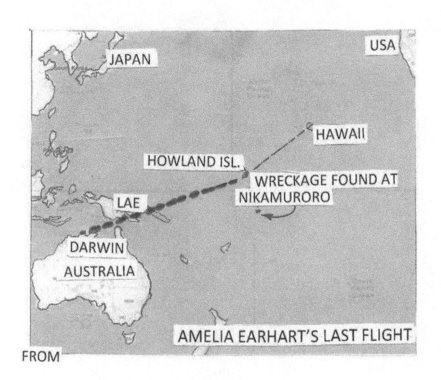

FROM

AMELIA EARHART'S LAST FLIGHT

Wiley Post's Russian Route

CHAPTER 5
MARVELOUS MOSSIE

Pilots called it Mossie. The de Haviland DH98 *Mosquito* was born in November 1940 and eventually flown by the air forces of 14 countries. A total of 7,781 were built, and there were more than 40 versions of the design. Its conception and gestation periods were somewhat unusual, dating back to 1936 when the British Air Ministry issued Specification P.13/36 for a twin-engine bomber. Its speed was to be 275 mph at 15,000 feet, and it had to have a range of 3,000 miles with a 4,000-pound bombload. In 1940, Geoffrey de Havilland sent a letter to Air Marshall Sir Richard Freeman, a senior defense official, suggesting that the airplane be made of wood and that DH could produce a bomber that would be so fast that the government's specified defensive armament would not be required. In addition, the material was readily available, would not impact supplies of aluminum used by other aircraft, and could utilize the skills of companies such as furniture manufacturers who were anxious to play some part in the country's war effort. He noted that wood resulted in a smooth rivet-free surface and was easy to repair. De Havilland contemplated a streamlined body with no draggy gun turrets. Construction would be based on a monocoque design—no interior bracing required—with skins made from a sandwich of plywood with a balsa core.

The government thought little of DH's proposal, so de Havilland decided to continue on their own. They quickly concluded that a bomber could be designed that was faster than German fighters, thereby confirming the notion that gun turrets, along with their crew and equipment, were not needed. This would allow only two crew members. Two Rolls-Royce Merlin engines were to be used, and they predicted the airplane would be capable of 388 mph maximum speed—23 mph faster than the *Spitfire*. Naturally, the idea of only two crew members in a bomber and speeds faster than their top fighter was somewhat hard for the Air Ministry to believe.

For secrecy from hostile nations and the prying eyes of their own government, the entire design team was moved to a 300-year-old country house a few miles from their base at Hatfield. The location was Salisbury Hall, just south of the old city of St. Albans, a short train ride north of London. It was where legendary Nell Gwyn lived while mistress to King Charles II, and a moat surrounded it. This latter feature proved to be quite a hazard because a swinging rope bridge was used to cross the moat to reach the workshops and canteen on the other side. Consequently, employees had to use this, sometimes with an armful of drawings where some didn't make it. They get dropped into the moat! I remember crossing it several times per day for twelve months and have vivid recollections of the "bridge" whipping up and down caused by another person jumping on it while I was halfway across). The area surrounding the house was camouflaged, including the cars, to maintain its rural appearance, and no bombing attacks were made on it. However, bombs dropped in the surrounding area. And in one case, a magnetic mine parachuted down but was caught up in the trees. If it hadn't been for those trees, the entire project would have been wiped out.

Despite Air Ministry opposition, Freeman had enough confidence in the DH proposal to get a specification issued (B.1/40) to match the preliminary design, but only for one unarmed reconnaissance prototype, which kept the program alive. The Ministry of Aircraft Production, headed by Lord Beaverbrook, tried several times to stop work on the aircraft but later agreed to have DH produce 50 aircraft by July 1941.

R. E. Bishop headed the original design team of nine men and worked 11 hours per day for six days, plus 4 hours on Sundays. The decision to locate away from the main plant at Hatfield turned out to be wise since work at Hatfield was constantly interrupted by air raids. In late 1940, for example, bombs dropped within a mile of the factory several days per week. In one case, the plant was bombed, killing 21 people and destroying most of the de Havilland Aeronautical Technical School. Its students were moved to Salisbury Hall after the Mosquito design team left.

On November 3, 1941 the Mosquito prototype was split into two sections—fuselage and wing. This was taken by truck six miles away to Hatfield, then reassembled. Engine runs began, after which young Geoffrey de Havilland Jr., chief test pilot, did some taxi trials and made the first flight. The design started in December 1939. Its first flight was on November 25, 1940, remarkably in just 11 months from its start, and within another 11 months, it was operational! Subsequently, the prototype was used henceforth to check the aerodynamics, experiment with changes, and for demonstration. About a month after its first flight, Geoffrey put on one of his dramatic flying displays with it in front of Lord Beaverbrook and other senior officials. They were astounded and immediately increased the order from 50 to 200. Development work proceeded on four variants, the original reconnaissance version, a fighter, a dual-control trainer for operational training, and a bomber, labeled Marks I, II, III, and IV, respectively.

So, let's see what made the Mossie so special. It is worth noting that the first Mosquitoes were hand-built in a small hangar behind the house and across the moat!. A second hangar was added when production ramped up. All of the jigs that they madewere inside of this hangar. All of the machining was done there, and they even built a special landing gear for the airplane that didn't require the expensive high-precision parts used today. Instead, they used rubber blocks stacked on top of each other in an oval cross-section tube as the shock absorber.

The fuselage was split into left and right halves, enabling simple installation of parts. Each half was made by stretching three thin diagonal layers of birch plywood over a concrete mold into which the bulkheads and other structural pieces had been inserted.

Essentially, this mold was a concrete model of a complete half fuselage. Double curvature was obtained by using straps to stretch the plywood into shape. Blocks of balsa wood, 3/8 inches thick, were glued onto the inside layer of plywood, and then a final layer of birch plywood was glued and stretched into position. The inner and outer layers of plywood each measured about 5mm (0.20 inches), so total skin thickness was about ¾ inches. The upper and lower edges, where the left and right fuselage sides were joined together, were made of male and female spruce wedges glued and screwed to join the halves.

The upper wing skin used a double plywood sandwich with square-section stringers made from Douglas fir, and the lower skin was made from a single plywood panel with spruce stringers. Wing spars were made in one long tip-to-tip piece with laminated spruce booms boxed with plywood webs. This choice of materials and construction resulted from de Havilland's long experience with wood and from a series of panel tests. It is worth noting that it wasn't just "any old spruce tree" that would suffice. Only ten percent of Canadian spruce trees met the requirements for long straight grain, moisture, density and so on. A high degree of accuracy was achieved—sawing was within 0.01 inches and the 50 feet long wing spar was accurate to within 0.04 inches. At first, these parts were glued together using casein glue but later changed to a synthetic glue, known as "beetle," to eliminate fungus growth. When a spar, for example, was being glued together, the glue was applied while the parts were still in their jigs. Then the pressure was applied, and the whole thing was covered with electric mats to provide a constant temperature. In some cases, the joint would be further reinforced with screws. Interestingly, tests showed that the wood failed before the glue when loads were applied. Where cutouts were required, the edges were reinforced with ash. Wing ribs had laminated spruce upper caps, Douglas fir lower caps, and three-ply webs. Walnut was used in the area where the fuselage is joined to the wing since it is a strong hardwood that distributes the bolt loads at that joint. Most of the control surfaces, such as flaps, were fabric-covered. I also recollect the entire body is covered with a layer of doped-on fabric to give it a smooth surface throughout.

The aircraft proved to be remarkably strong in service and capable of absorbing considerable battle damage. When damage had to be repaired, carpenters merely cut away the edges, smoothed them out, and glued in some new wood as required, using screws as needed.

Some of the "old-bomber guys" in Whitehall still thought a gun turret should be added, so DH added one and tested it. As predicted, the extra drag cut the speed by 20 mph, and that idea fizzled out. The bomber version made its first flight in September 1941, carrying four 500-pound bombs. The highest altitude ever achieved was 40,000 feet, and its highest speed ever was 437 mph. The fastest in the world until the jets came along. Its highest bomb load was a 4,000-pound "Cookie," a load equivalent to that of the B-17 "Flying Fortress." And to do this, the bomb bay was enlarged with a bulge beneath. We called it "the pregnant Mossie."

The *Mosquito* was used in many roles—night fighter, fighter-bomber, a pathfinder for the big heavy-bomber raids, ship-sinker, photo- reconnaissance, high-speed courier, and so on. They were making nightly raids on Berlin for a long time, much to the dismay of Luftwaffe chief Herman Goering who said that their capital city would never be bombed. He said, "I turn yellow and green with envy when I see the Mosquito. The British knock together a beautiful wooden aircraft that every piano factory over there is making. They have the geniuses; I have the nincompoops."

Their specialty was sudden pin-point bombing. Laser-guided and GPS bombing had not been invented back then, and conventional bombing was inaccurate. The Mossie, however, would swoop in across the treetops, drop its bombs right on target and dart out again before the enemy knew what was happening. Gestapo headquarters buildings were destroyed, with only minor civilian casualties or collateral damage in Oslo, Copenhagen and the Hague to destroy records and save the lives of many resistance fighters.

One particular raid is noteworthy—the attack on Amiens prison in northern France. The French resistance fighters had radioed London saying that 12 of their people had been shot at Amiens and that about 100 more were listed to be shot too. The prison was essentially escape- proof, and they pleaded for help. The Royal Air Force responded by planning an air attack. The prison was crucifix-shaped with a separate building for the guards, and a 20-feet-high wall, 3-feet thick, surrounded the entire complex. It was obvious that the attack would have to be very precise. For example, they would need just enough explosives to break a hole in the wall and open the cell doors. Too little explosive would accomplish nothing, and too much would cause the deaths of those prisoners they were trying to save. Nine *Mosquitoes* were used in the attack, which began about noon on February 18, 1944. The first three aircraft zoomed in and scored a direct hit on the guardhouse, killing or wounding all of those inside having their lunch. The second group flew in and finished the job. They could see the prisoners streaming out through the wall. One Mossie was lost in a crash landing nearby, killing the navigator, and another was lost when an FW 190 shot its tail off, killing both crew members. As a result of the raid, 255 prisoners escaped and 87 prison occupants were killed, including German guards.

In a completely different type of operation, a Royal Canadian Air Force Mossie fighter version was flying over the Baltic coast one spring day in 1944 when they noticed a Heinkel He 111. They chased it and shot it down, and right after that they saw a Focke-Wulfe 190 nearby. They had a brief dogfight with it before shooting it down too. Then they saw a line of aircraft sitting on a runway, but there some enemy aircraft flew above them. They destroyed two of them and then shot up the ones on the ground. To wrap things up, they shot down a seaplane and returned home. Air Marshall Goering was becoming very upset by these *Mosquitoes* wreaking havoc. He distributed a poster to German air bases offering a reward to any Luftwaffe pilot who destroyed one.

As noted earlier, one version carried a 4,000-pound bomb, and in one memorable attack, they dropped one from a low level while trying to block a vital rail tunnel. It went through the tunnel entrance, scooted along the tracks inside, and then exploded! They were also used extensively in attacks on Berlin, including a two-night raid involving eight squadrons of *Mosquitoes*, and only one aircraft was lost in these two nights.

Another version was equipped with a 57-mm cannon used primarily to destroy ships. It was equipped with extra fuel tanks to improve the range, allowing it to mosey around looking for U-boats in the Bay of Biscay. It also had four machine guns to take care of any enemy aircraft escorting the shipping. Not only were quite a few vessels destroyed or badly damaged, but it required the Germans to divert other valuable air and sea resources to provide escorts to the surface ships. Later, the aircraft was equipped with four racks on each wing to fire rockets. These proved to be very effective when shot close to the ship's waterline allowing water to enter the ship and causing fires which ultimately caused it to sink.

The photo-reconnaissance versions of rockets swept the North Sea and Channel coastline and discovered the V-1 flying bomb launchers and the highly-secret V-2 rocket sites at Peenemunde. Bombers subsequently attacked both. Over 2,000 V-1s were launched against England, and their small size made them difficult targets. They were also fast and flew at low altitudes, so fighters were, in most cases, unable to stop them. When their engine cut out, they plummeted to the ground creating considerable damage. So it was imperative to destroy the launch sites. The V-2 rocket was essentially an early version of today's ICBM, and at that time, it was impossible to destroy once it had been launched. Some 600 allied bombers on each of two nights bombed Peenemunde, and this so severely damaged the V-2 program that they were never able to make it the menace they intended. A number of them were launched, but they seemed to have zero accuracies, and when they did hit the ground, we'd just shrug our shoulders and say, "Sounds like another V-2, but we're OK."

The night fighter role was a great success, primarily due to the advent of airborne radar. The Germans were puzzled about how Wing Commander John Cunningham could see them so well in the dark. The British were anxious to keep their radar secret so they tried to convince everyone that his eyesight at night was enhanced by eating lots of carrots! To understand the technique, this is what happened. An unbroken chain of tall radar towers detects the aircraft while they are a long way from the coast. This information is transmitted to the control center for that area, and appropriate pucks are moved into position on the huge table. When the aircraft crosses land, it is picked up by the Royal Observer Corps (ROC). This is a network of "posts" manned by civilian volunteers such as mailmen, farm laborers, pensioners and shopkeepers who work in pairs on four-hour shifts every day with posts scattered across the entire country, a few miles apart. Each post has a device they use to determine the airplane's altitude and location within a mile. When they detect an airplane, they contact the control and say they have an airplane, giving its type, such as Ju 88, if they can see it, altitude and "over square number so-and-so." In one case, where my Dad was the head observer, they cross-plotted at night with two other posts and, using their ears, were able to locate an enemy bomber precisely. The control center passed this information to a night fighter which dropped down through a cloud and saw the enemy right where it was supposed to be. It was appropriately taken care of.

Although the low-level, high-speed attacks were very successful overall and demoralized the enemy, they were hazardous. Many *Mosquitoes* were lost due to them hitting obstacles. One, for example, hit a flagpole in one of the Gestapo HQ raids. Some got a wee bit too close to the wave caps as they flew over the North Sea and Channel and ended up floating, and yet another flew home with a mass of power cables hanging from the wing—and the copper was recycled!

Finally, to satisfy those who have fond memories of Olivia de Havilland, the beautiful actress who won two Academy Awards, she and her sister Joan Fontaine were cousins of Sir Geoffrey. "Young" Geoffrey, the test pilot, visited them in Los Angeles while demonstrating the *Mosquito* to U.S. personnel during the war.

de Havilland Mosquito

CHAPTER 6
THE SPITFIRE/MUSTANG STORY

In February 1935 the Messerschmitt Me 109 was intoduced into Germany's Luftwaffe. In Britain, the RAF introduced the Hawker *Hurricane* in December 1937, and those two were the world's top two fighters when war clouds began to heave up. Most of the other fighters were biplanes, such as the Hawker Fury and Gladiator.

Then came the *Spitfire*. While Messerschmitt was designing the Me 109, Supermarine Aviation Works, a subsidiary of Vickers-Armstrong Ltd., developed Schneider Cup racing seaplanes. Notably, the S.6B won the trophy outright for Great Britain after winning the race three times in a row. It had an average speed of 340 mph. Later, with a more powerful engine, an S.6B broke the world's speed record by achieving 408 mph. Supermarine's chief designer, Reginald J. Mitchell, then turned his attention to land planes in order to satisfy the requirements of Air Ministry specification F7/30 for a new 250 mph fighter.

After several iterations, Supermarine developed their Type 300, later called the *Spitfire*, and a revised specification (F10/35) was prepared to match the characteristics of that airplane. It had a beautifully designed streamlined shape with elliptical ultra-thin wings and was powered by a new Rolls-Royce V-12 engine, the *Merlin*. Its armament was four0.303 Browning machine guns in the wings, and like the Me 109, it had narrow-track retractable landing gear.

The *Spitfire* (K 5054) made its first flight in March 5, 1936, from Eastleigh Aerodrome (Southampton).

Following the f light there were the inevitable changes such as a new propeller, a later engine model, control improvements etc. This substantially improved the aircraft so that the government issued a contract in June 1936 for 310 aircraft, which created a problem because the relatively small Supermarine factory was already working at full capacity. Eventually they found ways to overcome that bottleneck, but it was not until mid-May 1938 that the first production aircraft flew.

War clouds were appearing on the horizon, and I can't help thinking how beneficial it was that the much-maligned Neville Chamberlain was able to stave off the war for a year. When I think of how few modern fighters were in the RAF in 1938, it is obvious that a war at would have ended differently to the way it did in 1945. The U.S. was in no position either to influence the outcome because, like Great Britain, they had decided to ignore the looming menace of Nazi Germany. In the nick of time, the British government woke up and paid attention to the problem. In addition to ordering a sizeable number of new fighters, they also realized how vulnerable the factories were to air attacks. Consequently, they decided to build some "shadow factories." For the *Spitfire* they built one near Birmingham, and this proved to be fortuitous when the main plant was destroyed in a bombing raid in September 1941. The new plant, Castle Bromwich, had its growing pains; labor problems and the difficult-to-mass produce complex shapes of the airplane, with its stressed-metal skin and elliptical wing, led to the factory being taken over by the government. Eventually, it became the largest *Spitfire* factory, turning out a maximum of 320 aircraft per month and reaching 20,000 aircraft. It was produced in greater numbers than any other British aircraft and saw service with many air forces, including the U.S. Army Air Corps, which had 600 of them.

R. J. Mitchell didn't live long enough to see the great success of his brainchild. He had cancer but lived long enough to see its first flight. He died in 1937 at age 42. The airplane's development was taken over by Joseph Smith, who had been chief draftsman during Mitchell's reign.

The *Spitfire's* first significant action was the Battle of Britain that started officially on July 10, 1940 and lasted until October 2, although the major battles were fought between August 2 and September 24. It was accompanied by the *Hurricane*—a less sophisticated airplane, using a braced metal fuselage skeleton, wood frames and a doped fabric skin that was easy to repair. It was not as fast as the Spit but could out- turn both the Spit and the Me 109. There were more *Hurricanes* than *Spitfires* available, so it wasn't surprising that they downed 55 percent of the enemy aircraft. It was also found that the Me 109's cannon shells would go right through the fabric skin without exploding – a distinct advantage! The Spit also encountered a problem in these early battles that were quickly corrected. It had an old-fashioned float-type carburetor that cut off fuel to the engine whenever the airplane went into a negative-G maneuver. Although the engine started again shortly after, the enemy aircraft usually able escaped in the interval. The net result was that the *Spitfire* had a lower attrition rate and a higher victory- to-loss ratio than the *Hurricane*, and for various reasons, they won the battle. The British were more capable than the Germans in replacing lost or damaged aircraft and pilots, and also, the short-range Me 109s frequently had to abandon the battle to have enough fuel left to fly home. This reason had another side-effect. If a German pilot couldn't make it home, he either ditched in the Channel or crash-landed in England. In both cases, the pilot was then lost to the Germans. In contrast, the RAF pilots were able to parachute into friendly territory or crash-land close to their shore or in England where, in both events, they would be available or duty again if and when they were alive and well.

Twenty-four variants, called Marks, of *Spitfires*, were produced, including fighters, fighter-bombers, photo-reconnaissance, and even carrier types. They were a favorite of the pilots and were widely used throughout the war. In the defense of Malta, for example, they were operated from the deck of the U. S. Navy aircraft carrier U.S.S. *Wasp* to provide substantial assistance. They were used in the North Africa campaigns. The most popular version was the Mk Vb, with 6,479 of it built. It had a maximum speed of 378 mph, a combat radius of 470 miles and an armament comprising two 20mm cannons plus four .303 machine guns and two 240 lb. bombs—a substantial improvement over its predecessors.

The *Mustang*, more properly known as the North American P-51, was conceived as a result of a British requirement, not as an American one as most people would think. In early 1940, about two years before the U.S. entered the war, the British Air Purchasing Commission in New York looked for a fighter to quickly supplement their air force. Great Britain already had its factories working full steam, and they wanted to take advantage of America's untapped production capability. The Curtiss P-40 *Tomahawk* was the only U.S. fighter that met their requirements, but Curtiss was overloaded with work, and so North American Aviation (NAA) was asked if they would make P-40s for them. Instead, NAA's president, James "Dutch" Kindelberger, suggested a new airplane. That interested the British, but they said that it would have to be ready within the 120 days they requested. Kindelberger agreed to that and he did it in 117 days! Edgar Schmued headed the design team, and the NA-73 prototype, as it was called then, made its first flight on October 26, 1940, with the first production aircraft coming off the line on May 1, 1941. Like the *Spitfire*, it had a monocoque construction with aluminum alloy skins, leading to an initial contract for 320 aircraft. It was subsequently modified and developed over the following few years to become one of the best airplanes of World War II. Its main attribute was its long range. The *Spitfire* could fly as far as northern France and over Belgium and the Netherlands but could not reach much beyond the German border. Consequently, it could not provide adequate protection for bombers on their way to Berlin, Hanover, or the major industrial centers down the Rhine Valley. That was where the *Mustang* came in with its long range of more than 1,000 miles.

It fulfilled several roles as a basic fighter, a ground-attack aircraft, and a dive bomber. As a fighter, it had either eight 0.50-inch machine guns or four 20 mm. cannons. For ground attack or dive-bombing, it would have six 0.50-inch machine guns plus 1,000 lb. of bombs beneath the wings.

Early in its life, it became apparent that the original Allison engine was woefully inadequate at higher altitudes, so it was replaced with a Rolls-Royce *Merlin* engine built by Packard. This proved to be the master-stroke that changed the airplane into what many considered the world's best fighter. It was as fast as any other, could operate well at high altitudes, and had a long range. The only deficiency was rearward visibility, and they fixed that by using a bubble-top canopy.

The RAF initially used Allison-powered version in frequent sweeps of enemy territory in northern France and the low countries, including such "fun activities" as train-busting. In October 1942, they were the first England-based single-engine fighters to actually raid Germany when they attacked the Dortmund- Ems Canal. They were also used effectively in the 1943 offensive in the Mediterranean area, including the attacks against Sicily and Italy.

By this time, the P-51 was breezing along at 432 mph at 22,000 feet, and the U.S. Army Air Force (USAAF) had ordered more than 2,000 of them.

In 1942, the USAAF thought that their tightly-packed formations of heavily-armed bombers would not need fighter escorts, and initially, they were convinced that they were right. The formations of B-17s and B-24s were getting through to their targets with a loss of less than 2%. But when the Luftwaffe began moving more aircraft from their eastern front to counteract these raids, the losses mounted. In the Schweinfurt- Regensburg mission in August 1943, they lost 60 B-17s from a force of 376, and two months later they lost 77 out of 291 aircraft. They considered using the Lockheed P-38 *Lightnings* and Republic P-47

Thunderbolts as escorts but decided that the *Mustang* would be the best if it were equipped with external fuel tanks. At first, they used a combination of all three escorts but later settled on *Mustangs* for the entire mission. The Fw 190s had poor performance at the altitudes at which the B-17s operated, and the Me 109G was the only formidable opposition they had to encounter. Consequently, the bombers were now able to get through and wreak havoc down below. Some of the ensuing air battles were tremendous. One account describes 800 fighters escorting 1300 bombers meeting a thousand Luftwaffe defenders! In another case, we read of *Mustangs* escorting the bombers all the way to Berlin, 1,100 miles from their home base in England.

After the war, both the *Spitfire* and the *Mustang* continued to be used by the air forces of several countries until the 1960s, and it is not unusual to find *Mustangs* popping up at air shows even today.

	Spitfire Mk. VB	Mustang P51D
Maximum takeoff weight	6,700lb	12,100 lb
Maximum speed	370 mph	437 mph
Combat radius	470 miles	825 miles *
Service ceiling	36,500feet	41,900 feet
Rate of climb	2,600fpm	3,200 fpm

*Range with external tanks: 1,650 miles

Supermarine Spitfire

North American P-51 Mustang

CHAPTER 7
LOCKHEED & THE SKUNK WORKS

Before drawing back the curtain on the secretive Skunk Works, it is interesting to explore the beginnings of Lockheed, the largest defense contractor in the U.S., and the first unusual item, which is its name. There never was a Mister Lockheed involved! Instead, the name came from the Lougheed brothers Victor, Allan, born in 1887, and Malcolm, born in 1889. It is an Irish name usually pronounced as "Lougheed." Eventually, tiring of the confusion, an early 1919 poster for Loughead Aircraft Manufacturing Company suggested that it be pronounced "Lockheed,". In 1921, Malcolm used Lockheed for his hydraulic brake company in Detroit for the first time.

When the three boys left their home near San Francisco, Victor went to Chicago to work on automobile engineering; Allan went to work in a repair shop, and Malcolm became an engine mechanic on steam cars. Later, Allan became a mechanic on a Curtiss pusher biplane owned by Malcolm's boss. He taught himself to fly and occasionally flew as a copilot on the biplane. When a problem occurred one day, and the regular pilots could not make it take off, Allan volunteered to do it, which he did! He made some adjustments to the rigging and the engine and was able to take off on his second attempt. He was, of course, bubbling over with confidence by this turn of events and stayed there for about a year as a mechanic and pilot, after which he moved to the International Aviation Company of Chicago as a flying instructor and exhibition pilot. In 1912, Allan returned to San Francisco, where he and Malcolm decided to design and build a seaplane, working as mechanics during the day and doing their airplane at nights and weekends.

The seaplane, which they called Model G for no apparent reason, was a biplane with a 46-feet wing span and a 30-feet-long fuselage. They installed an 80 hp Curtiss OX engine driving a tractor propeller. Allan took it on its first flight on June 15, 1913. Two years later, at the Panama-Pacific Exposition in San Francisco, they made $6,000 flying passengers on short hops. Emboldened by their success, they formed the Loughead Aircraft Manufacturing Company, with its workshop close to the waterfront in Santa Monica, California. They hired a young architectural draftsman who'd taught himself the fundamentals of stress analysis to help them. His name was Jack Northrop, who later became famous as the Vega designer, and later still founded Northrop Aircraft in Hawthorne, California. Their new project was a twin-engine, 10-seat, flying boat, the F-1, which was a success. They tried to sell it to the U.S. Navy, but the Navy had already chosen the Curtiss HS-2L to fulfill its needs. They designed a landplane version and managed to make money by renting it out to movie companies for aerial footage.

Their next project was a small seaplane, the S-1. It had a 21-feet-long wooden fuselage built in two symmetrical halves formed in concrete shells. These two halves were made of laminated spruce veneers with grains at right angles to each other, and pressure was applied to the glued assembly by inflated bags. They were then attached to a framework of frames and bulkheads. Its upper and lower wings were fabric-covered, and a 25-hp two-cylinder engine provided power. However, there were many inexpensive war-surplus aircraft at that time, so Loughead could not sell any S-1s, and the company was dissolved in 1931.

Malcolm turned his attention to a four-wheel hydraulic brake system that he'd invented for automobiles. As noted earlier, he decided to use "Lockheed" when he founded the Lockheed Hydraulic Brake Company in 1919. After several successful years, he eventually sold his company in 1932.

Allan acted as an agent for Malcolm's brakes, and Jack Northrop went to work for Donald Douglas in Santa Monica. His boss at Douglas was "Dutch" Kindelberger. In his spare time, Northrop worked on a single-engine transport with an S-1 type construction, but capable of carrying a pilot and four passengers. He showed his design to Allan who was still anxious to return to the airplane business but they needed money. They found a wealthy brick and tile manufacturer named Fred Keeler, who agreed to form a company with him as president and Allan as vice-president. Northrop was THE engineer, and the company was called LOCKHEED AIRCRAFT COMPANY, located in Hollywood, California.

The aircraft that Jack Northrop had envisaged became known as the Vega, and the previous descriptions of Wiley Post's and Amelia Earhart's exploits show how the Vega became world-famous.

The 1927 Vega was a masterpiece—a streamlined, high-performance, high-wing monoplane employing a monocoque wooden fuselage. It made its first flight on July 4, 1927 from Mines Field, now Los Angeles International Airport. It was called *The Golden Eagle* and disappeared without a trace while flying to Hawaii in the Dole Race on August 4, 1927. Hubert Wilkins, later Sir Hubert, an Australian explorer, bought a subsequent model. He used it to fly over various parts of the Arctic after equipping it with skis and extra fuel tanks. On April 15, 1928, he and his friend Ben Alison flew it from Point Barrow, Alaska, to Spitsbergen, a distance of 2,200 miles, where a blizzard was blowing. Later that year, Wilkins embarked on a series of flights, using two Vegas, the *Los* Angeles and the *San Francisco*. Starting at the South Shetland Islands, they discovered and named various areas, such as *Cape Northrop* and *Lockheed Mountains*, in honor of you-know-who.

With all of the attendant publicity, Lockheed's order books were bulging, and they needed more manufacturing space, so they moved from Hollywood to Burbank in March 1928. They used a building partly occupied by the Mission Glass Works, and outside there were vineyards, farms, orchards and miles of desert. To west, the area became the United Airport by 1928 and eventually Burbank-Glendale-Pasadena Airport.

In 1928, Western Air Express talked to Lockheed about a Vega version with more power and higher speed. This was accomplished by doubling the engine's power, using a Pratt & Whitney Wasp engine instead of the Wright Whirlwind. The first of these was named *Yankee Doodle, and* it became the first aircraft to fly non-stop across the U.S. in less than 24 hours.

Jack Northrop left Lockheed in June 1928 and was replaced by Gerard Vultee, who also founded his own airplane company later. Before leaving Lockheed, Northrop designed an airliner version of the *Wasp*- powered *Vega* for Western Air Express, but this had an open cockpit to satisfy their wishes. Airmail pilots preferred such an arrangement, apparently. It was called the *Air Express*, and it was 35 mph faster than the *Vega*. Vultee, as the new Chief Engineer, worked on the subsequent development of the *Air Express* and Lockheed's reputation was as high as its order books at that time. An auspicious time for the board to sell the company to the Detroit Aircraft Corporation, a holding company. Allan Loughead was not in favor of this transaction, but he was outvoted. He did, however, agree to be vice-president, but this essentially ended his aeronautical career.

So, in July 1929, Lockheed became a subsidiary of Detroit Aircraft and it continued to prosper. The *Vega* broke speed records of up to 340 mph and the altitude records of 55,000 feet. Low wing models were studied, and the *Sirius* open-cockpit version was notable. It was designed by Vultee to meet a specification from Charles Lindbergh. Subsequently, Lindbergh and his wife Anne flew it in various parts of the world, using floats instead of landing gear for some flights. One report estimates that they covered 30,000 miles over five continents while surveying future airline routes.

The next project was the *Altair*, a low-wing monoplane with hydraulically- powered retractable landing gear, the first U.S. military aircraft to have such a gear. The most famous *Altair* was Sir Charles Kingsford-Smith's *Lady Southern Cross* which he flew from Australia to the U.S. by way of Fiji and Hawaii.

The *Orion* came next. A good-looking, low-wing airplane with an enclosed cockpit, a six-passenger cabin, and retractable landing gear. It had a top speed of 230 mph, and two of these were bought by Swissair. It was the first American transport to be exported to Europe, and with its bright red paint scheme, it was widely admired.

Fighter aircraft in the world's air forces at that time were biplanes with fixed landing gears and open cockpits. So, Lockheed decided to make a great leap forward by using its expertise in fast, low-wing, highly maneuverable aircraft. The result was their first fighter—the XP-900. It had an all-metal fuselage and wooden wing and a 600-hp Curtiss V-1570-33 engine. For armament, it had two machine guns firing through the propeller and a single machine gun operated by a gunner in the aft cockpit. In September 1931, it made its first flight and demonstrated capabilities superior to other fighters. Its general arrangement was a significant step forward and quite similar to the famous fighters of World War II. However, work ended—except for some subsequent development by Consolidated Aircraft—when the Stock Market crash put ended Detroit Aircraft. Lockheed was still profitable, but other parts of that company were hit hard, resulting in Detroit's stock plunging from $15 per share to 12 ½ cents, sending it into receivership.

In June 1932, the assets of Lockheed were sold to three men in Los Angeles for $40,000. They were Robert E. Gross, Carl B. Squier and Lloyd C. Stearman. The story of how these men got together is quite interesting and quite complicated. So, in order to keep this overall review running smoothly, we'll just say that Gross had been an investment banker who had become involved with aviation when he and his younger brother, Courtland, started a flying boat company. He met Stearman during his investment business. Stearman was well known as the designer and builder of airplanes. Squier had been the general manager at Lockheed and had experience with Stinson and Glenn Martin before. It was he who proposed the idea of buying Lockheed for $40,000.

In the new company, thereafter known as the LOCKHEED AIRCRAFT CORPORATION, Stearman was president and general manager, Squier was vice-president, and Gross was treasurer.

When considering their next step, they concluded that single-engine transports would not meet safety requirements for regular operation at night or over areas not suitable for emergency landings, the Rocky Mountains, for example, so they began considering twin-engine aircraft, just as Boeing and Douglas were doing. The Model 10 *Electra* was the result, and this is the point where Kelly Johnson enters the picture. While studying aeronautical engineering at the University of Michigan in March 1933, at age 23, he participated in a wind tunnel test of the *Electra*. He found problems with its longitudinal stability and yaw control. Six weeks later, having now graduated with a Master of Science, he joined Lockheed. His first job was to correct the problems he'd found with the *Electra*. The airplane made its first flight on February 23, 1934. The airplane had a 55-feet wing span and was nearly 36 feet long. It was powered by two Pratt & Whitney R985-SB engines of 450 hp and had a maximum speed of 202 mph.

Lockheed then produced a smaller version of the Model 10, called the Model 12. It sold well, and this time it included a military version for the Netherlands. Some were sold with commercial markings to the British and French to reconnaissance over Germany prior to World War II.

Kelly then developed the Model 14 *Super Electra*, and its first flight was in 1937. It also found many customers, mostly foreign. In one of those, Howard Hughes made a record-breaking flight around the world in 1938. He did it in three days, 17 hours, and 14 minutes. It was also the airplane used to carry British Prime Minister Neville Chamberlain to his 1938 meetings with Adolf Hitler in attempts to prevent war.

From that point onwards there were the Hudsons, Lodestar, Ventura, the elegant Constellation, the twin-boom Lightning, known as the Fork-Tailed Devil by the Luftwaffe, and then the F-80 *Shooting Star.* This brings us to the **Skunk Works.**

General Henry "Hap" Arnold, U.S. Chief of Staff, was bothered that the British and Germans were all flying jet-propelled fighters by the early 1940s, while the U.S. didn't even have one on the drawing boards. So, he was delighted to hear that Lockheed's top engineers, Hall Hibbard, Kelly Johnson, and Willis Hawkins, had been considering it for some time. On June 23, 1943, the USAAF issued a contract for a fighter to be ready in 180 days. Using the British de Havilland *Goblin* jet engine, a team of 23 engineers and 105 shopmen were put to work in a temporary wood and canvas building to maintain secrecy. And this became the first home of Lockheed's Advanced Development Projects (ADP) division, known generally as the Skunk Works.

Where did that name come from? There have been several explanations, but the most likely is given in Kelly Johnson's autobiography. One of his engineers had been asked what he was doing, and he was told that Johnson was "stirring up some kind of brew." In the comic strip *Lil' Abner*, a character was regularly stirring up a foul-smelling brew called *Kickapoo Joy Juice* that contained live skunks as a primary ingredient, so somehow the term Skunk Works seemed to be appropriate. It is interesting to note that since then, the name has sometimes been applied by other companies to any area where secret work is underway.

Working 60-hour weeks, the prototype XP-80 was accepted by the USAF on November 15, 1943, and within two months made its first flight. It was a great success, followed by a second prototype that was powered by a General Electric J33 engine based on the *Goblin*. The USAAF ordered 13 of these, and soon afterwards, 5,000. By early 1945, two of them were sent to England hoping that they'd have a chance to duel against some Messerschmitt Me 262s, but to no avail. The war ended before the P-80 could prove itself against the Luftwaffe. When the U.S. Army Air Corps became the U.S. Air Force (USAF) in 1947, the designation P for "Pursuit" was changed to F for "Fighter." The P-80 was consequently changed to F-80, and at about that time an improved version, F-80B, was introduced with an ejection seat, better machine guns, and more power. Then came the F-80C with more powerful engines like the Allison J33, and they were the ones the USAF used in the Korean War. More than 900 were available when war began in June 1948. Two two days after it started, they shot down four Ilyushin bombers—their first "kill." After the Chinese entered the war, the F-80s faced the MiG-15 for the first time, and they shot down their first one in November 1950. But the later-technology Russian fighters were a tough foe, and the USAF began to replace the F-80 at that time with the new F-86 *Sabres* from North American.

Skunk Works is essentially a small autonomous division within a larger company. Its leader periodically presents progress reports to top management, and even then, the most secret portions may not be mentioned. Its leader has full authority over the budget, manpower, schedules, engineering, manufacturing, and testing. The benefit of being part of a larger company is that it can use that as a resource for manpower and equipment as required—using only specialists for the time they are needed. Security is vital. To quote Stephen Justice, heading up ADP at Lockheed, "It takes about one-tenth the time and one-tenth the resources to develop a countermeasure to anything that's introduced. To maintain your edge over any threat, you need to protect what your capabilities are. And sometimes, you need to protect their existence." Ben Rich, who was the General Manager of

ADP in 1988, said in a lecture to the Royal Aeronautical Society that "Good security begins and ends with keeping the organization lean. More people in the know means more people who do not really need to know." It essentially becomes a matter of pride to all personnel involved to observe tight security, and the overall arrangement of the facility is conducive to it. When I worked there, for example, I found that I could not access any areas other than the one related to our particular project. When a representative of a wheel and brake company came to talk to me about that part of my project, I had to talk to him in a conference room in the lobby, not adjacent to my desk, as would have been the normal case. A bare minimum of people are used, and shortcuts are made where possible. For example, an engineer may change a print (by "red-lining") in the workshop to make a quick fix and then go back to make it official by changing the actual drawing. Reports are kept to show thorough details of development, but they are kept to a minimum. For example, it was required to make thorough "trade studies" in order to determine a particular configuration, but in a Skunk Works environment, this could be resolved quickly and without reams of documentation by using common sense, instinct and background knowledge.

A major advantage of Skunk Works activity is the ability to take risks. As Ben Rich said in his lecture, "Without them [risks], you will not have successes, and successes and failure are the opposite sides of the same coin. That coin's name is risk." Skunk Works is small, and its costs are low, so it is not uncommon to find a small number of people working on a project that may only have a slim chance of succeeding, but if it did, it would be a major breakthrough or a big money-maker.

In the interview with *Air & Space Smithsonian*, Stephen Justice said that today's Skunk Works employs 1,700 people, working on more than 500 projects from radar coatings to war games to compact fusion reactors to a Mach 6 spy plane. That in itself shows how lean it is, and it is also obvious that a number of these will be failures or will be filed away for future development when the timing is right.

But this small size that is the essence of risk-taking—it allows many projects to be worked on and many to be rejected or later recognized as failures without any major impact on the company's finances.

So, it has been inevitable that the Skunk Works would have failures such as the *Saturn* small airliner and the XFV-1 *Pogo Stick* vertical riser that would have been able to blast off from any small area such as a ship's deck. However, these have all been overshadowed by its many successes, and the first of these, after the F-80, was the F-104 *Starfighter.* Design work on this started in November 1952, and within four months, the USAF ordered two prototypes. It was an immediate success. It first flew in February 1954 and, within a little more than a year, had flown at twice the speed of sound. Later, it flew at altitudes up to 91,249 feet, and many were sold both in the U.S. and elsewhere. They were also produced in factories in other countries.

In 1953 the USAF and CIA asked Lockheed to develop an airplane capable of flying over the Soviet Union at altitudes higher than any Soviet interceptors—a super-secret project ideally suited for the Skunk Works. This became the U-2. Work started in 1954, and by August 4, 1955, it was ready for its first flight. It was so secret that this flight was not made at Edwards Air Force Base but at an airfield built at Groom Lake in the Nevada desert. Within a year, it was making flights over Moscow and Leningrad. Until May 1960, the world became aware of its covert activities when Francis Gary Powers was shot down by a Soviet missile.

The next major Skunk Works project was the A-12/YF-12/SR-71. Lockheed received the contract in January 1960 for twelve A-12s. It first flew in April 1962. Essentially it began as a Mach 3 interceptor capable of operating at altitudes up to more than 95,000 feet and having a range of 2,500 miles. The A-12/YF-12 program was eventually terminated when its high cost and absence of Mach 3 threat were considered. However, work continued on a reconnaissance-strike version known later as the SR-71 *Blackbird.* In January 1966, the aircraft became operational with the U.S. Strategic Air Command, all of them painted with a special indigo blue radar-reflective paint. Hence, their name, *Blackbird*. The world became aware of it when it flew from New York to London in one hour and 56 minutes to participate in the 1974 Farnborough Air Show. Although its performance is classified, the information currently available shows its top speed at more than Mach 3.5, its operating altitude is more than 100,000 feet and its range at 3,250 miles.

Since then, Skunk Works has been deeply involved with stealth technology, particularly, the development of stealthy fighters such as the F-117 and the F-35. So, what's next? The aeronautical community would love to know. We know that a new bomber is being looked at, and undoubtedly there are space-flight projects, but Stephen Justice says that 500 projects are in work. I wonder what they are.

Lockheed F-104G

Lockheed Constellation

Lockheed C-5 Galaxy

Lockheed F-117

Lockheed F-35

LockheedMartin F-22

CHAPTER 8
THE COMET

There were two *Comets*; the first was the de Havilland DH 88 that first flew in September 1934. A twin-engine racing airplane that enabled C.W.A. Scott and Tom Campbell Black to win the MacRobertson Trophy in the London to Melbourne, Australia air race. When I last saw that airplane in the 1940s, it was gathering dust in the corner of a hangar at the Salisbury Hall. It has now been restored to flying condition.

The later *Comet*, the de Havilland DH 106, was the world's first jet airliner and is the airplane addressed at this point. It was born in July 1949 after a 3-year gestation period, and its beginnings were a result of the U.S./U.K. agreement in World War II, in which the U.S. was responsible for all transport designs during the war. The U.K. realized that their industry would have to catch up quickly to remain competitive in the post-war airline market. Their government initiated the Brabazon Committee that produced specifications for various airliner types, one of which was the Type IV that de Havilland responded to with the *Comet*. Work began in 1946, and British Overseas Airways Corporation (BOAC) was designated buyer.

I joined the DH stress office, where I was responsible for analyzing the structural integrity of wing flight controls and some parts of the inner wing. On one occasion, though, I was monitoring the strain gauges during pressure testing of the forward fuselage. If any of the strains deviated from a linear pattern, we had to shout out and stop the test. However, one day without warning from the pattern, there was a very loud bang inside the pressure chamber when the pressure was about 8 psi, as I remember it. Testing stopped, and the front of the chamber was removed. It showed the wood panelling inside the chamber in tatters, but the fuselage looked OK except for one place where the loads had concentrated down one windshield pillar and had ripped open a small area at the base of that pillar. It emphasized how such a relatively small fracture can have disastrous results in a pressurized cabin.

Despite the general attitude among airline operators that passenger jet travel would be uneconomical due to high fuel consumption, de Havilland thought otherwise, claiming that super high aerodynamic efficiency and high speed would result in a correspondingly higher number of passengers being carried in a given period. Subsequent route trials and operations proved DH was correct. It flew 70 percent further than a piston-engine plane between overhauls, thus resulting in lower costs per passenger mile.

Its aerodynamic efficiency was manifest in its sleek shape, minimizing drag by using Redux to bond stringers to skin rather than rivets, and by burying the four engines INSIDE the wing, i.e., they were not hanging in pods. By having engines inside the wing, they could be placed close to the wing root, thus reducing the yawing effect when one engine failed, which permitted a smaller vertical tail.

By current standards, the passenger load was small—48 passengers in the prototype, and 70 in the later (stretched) models. Cruising at about 40,000 feet, the passengers found it comfortable, free of engine vibration, relatively quiet, and very smooth. Its speed was another plus for the passengers since they got to their destination in about half the time of its predecessors.

The four engines used initially were the de Havilland *Ghost*, about 5,000 pounds thrust each, but they were replaced by the axial-flow Rolls-Royce *Avon* to provide more thrust and lower fuel consumption. These engines were mounted between the wing's front and rear spars, and this necessitated large holes in those spars for the intake ducts and tailpipes. Holes this large in such a vital part presented a problem, but the design proved OK and no issues were encountered. One interesting feature was the addition of a rocket engine for increased thrust when taking off from high-altitude hot runways, such as Nairobi. It was situated between the tailpipes of each pair of engines and provided 5,000 pounds thrust for nine seconds. This engine, known as the de Havilland *Sprite*, used hydrogen peroxide combined with permanganate as the propellant.

The airplane became operational in 1952 with BOAC and orders were received from many airlines but were suspended following two major accidents in January and April 1954 over the Mediterranean. Both were caused by metal fatigue, which was a poorly understood phenomenon at that time. Older airplanes had little, if any, pressurization in the cabin because they flew at much lower altitudes. When they developed a crack in their skin, it was usually of little consequence and was easily and quickly repaired. However, the *Comet* cruised at high altitude and had high pressurization. If you make a hole in a balloon, the result is catastrophic, which is what happened to the airplane.

Bits and pieces of the *Comet* were retrieved and meticulously put together by Britain's Royal Aircraft Establishment (RAE) at Farnborough. They also put a *Comet* in a huge water tank where they flexed the wings and pressurized the fuselage many times until failure occurred. The investigation concluded that fracture occurred near the direction- finding window above the cabin, and in a third of a second, the cabin would be completely emptied. Sir Arnold Hall, director of the RAE, said, "The terrific blast of air will force everything out of the hole and tend to throw the aircraft into violent conditions and tear it to pieces." It was also explained that metal fatigue was definitely the cause of the fracture. A structure with ample strength reserves when it is new might eventually fail.

As a result of this investigation, designers began to pay close heed to the so-called S/N Code, which relates stress to the number of operations and also to areas where stress concentration occurs, for example, at the corners of windows. The lessons learned were applied immediately to aircraft being designed on both sides of the Atlantic. I I know of at least one U.S. company that went so far as to immediately immerse its aircraft in a water tank to check it out and make appropriate modifications.

Boeing was racing to catch up with the *Comet* and had hoped to have its 707 as the first commercial jet airliner to cross the Atlantic. They failed to be the first, but the *Comet* disasters gave them the lead from that point onwards. The *Comet* was modified and stretched, and oval windows replaced the squarish ones used in the early model to avoid stress concentrations. *Comet* models 2, 3 and 4 were ordered by BOAC, Capital Airlines in the U.S., Aerolineas of Argentina, East African Airways, Olympic Airways, Mexicana, United Arab Airlines, Middle East Airlines, Sudan Airways, Kuwait Airways, the Royal Air Force, Malaysian Airways and Dan-Air.

Then in 1967, the *Nimrod* flew—a substantially modified *Comet*. It was designed for maritime patrol for the RAF and antti-submarine and maritime surveillance. It served very well in these roles until its retirement in 2011. So, the *Comet* was active in one role or another for nearly 60 years.

de Havilland Comet

CHAPTER 9
THE ABANDONED ARROW

It is hard to believe nowadays that a democratically-elected Prime Minister of a highly-regarded country could wield enough imperialistic power to essentially wipe out that country's aircraft industry in one day. Yet that is what happened in Canada in February 1959 when John Diefenbaker cancelled the Avro *Arrow* project, albeit with advice from a group of narrow-minded yes-men who couldn't, or wouldn't, understand the intricacies of the budget, the true nature of the threat against their nation, the complexity of appropriate defense, and the failure of a U.S. missile system that they thought would be adequate.

The Canadian aircraft industry had three major companies in the 1950s—Canadair in Montreal, plus de Havilland and Avro in Toronto. Canadair was not doing any design engineering of any consequence, and de Havilland was concentrating on their highly-acclaimed bush planes. Avro was the only company involved in designing and manufacturing sophisticated modern aircraft, and their engines, and it was, by far, the largest of these companies.

In early 1952 the Royal Canadian Air Force (RCAF) began thinking about replacing their CF-100 fighter. They concluded that they needed an airplane that had enough range to fly beyond the Distant Early Warning (DEW) line equipped with missiles and would be fast enough to destroy an aggressor before it could attack the highly populated areas just north of the U.S. border. The DEW line was a string of radar stations along the northern edge of Canada, i.e., above the Arctic Circle. Specifically, they wanted an airplane with a combat radius of 200 nautical miles, a 60,000-feet combat ceiling, a speed of Mach 2.0, a two-man crew, and all-weather capability. No aircraft anywhere in the world that could meet that requirement or on the drawing boards.

I was one of a small team in Avro's Initial Projects Office. A 15 feet x 30 feet room in the corner of the Engineering Department. We'd been working for two years or so on such an airplane, so when the call came, we were ready. The RCAF issued their requirements in March 1953, and Avro Canada responded quickly and was awarded a study contract by the middle of that year. Our small team grew rapidly, and the design gradually became a reality. The average age of those in the design department was 28, many of them young immigrants from the world's aircraft industries who'd never heard the words "it can't be done." Apart from the usual wind tunnel tests in 16 different models in both subsonic and supersonic tunnels, we had an unusual program that involving firing of large-scale models across Lake Ontario. At the high speeds involved, it was difficult, if not impossible, to obtain valid wind tunnel data that could predict aircraft controllability, so a number of 1/8 scale models were built with operating control surfaces and mechanisms to move them. Each of these models was mounted atop a huge Nike rocket booster and fired across the lake. After releasing from the booster, it sent back signals to show responses to control movements. This was a highly successful program. On at least one occasion, however, U.S. radar detected this object hurtling towards the U.S. and scrambled fighter jets to investigate! Many skeptics who doubted Avro's ability to design an airplane that could meet the extreme requirements, mostly because neither the U.S. nor British companies had been able to do so. The U.S. did try with their F-108, but that project was cancelled.

Many unusual features were included in the design; some of them quite controversial, and they can be attributed to the expertise of

Jim Chamberlin, the Canadian-born Chief Aerodynamicist, later Chief of Technical Design. For example, it had unheard-of reverse camber on its wing. It had a complete weapons package that could be lowered from the fuselage for loading and accommodated either *Falcon* or *Sparrow* missiles internally to minimize drag. A feature not included in U.S. fighters until many years later. It had artificial stability and a high wing to simplify engine installation and armament. Its overall radar/ weapons control was also far superior to other aircraft.

The *Arrow* made its first flight in March 1958 and flew supersonically on its third flight. On its seventh flight, it achieved Mach 1.52 at an altitude of 49,000 feet still accelerating and climbing. Within a few months, it became evident that it would meet all of the specification requirements and be a world-beater. Figures show that with the Orenda *Iroquois* engine installed, on a typical mission with droppable external fuel tanks, it would intercept an enemy aircraft nearly 500 miles away, engage in combat for five minutes at 50,000 feet, fly back to base, and be able to loiter for 15 minutes before landing. It would have flown at Mach 1.5 on its way to the target and cruised home at a leisurely Mach 0.91. An astounding performance at that time, and pretty damned good even by today's standards.

When it became evident that Canada was going to produce such an aircraft, the USAF became interested. The Canadian government encouraged them to participate because they realized that U.S. funds could be an important factor in the program's overall success. Essentially, the U.S. wanted to protect its ICBM sites in the northern midwest from any attack from Soviet aircraft, and at that time, it was felt that the Soviet Union could have a long-range supersonic bomber by the mid-1960s. With such an aircraft, it would be necessary to detect it while it was over the Arctic and destroy it within a few minutes after that. While a chain of anti-aircraft missiles could do the job, there was always the problem of such missiles shooting down a commercial airplane by mistake, hence the need for a piloted interceptor.

In mid-1954, 15 months after Avro started work on the *Arrow,* General Twining, Chief of Staff, USAF, asked the U.S. Air Research & Development Command (ARDC) to examine the *Arrow's* specification and the RCAF welcomed them with open arms. The ARDC was impressed, and a useful exchange of views followed, with, the U.S. agreeing to assist with wind tunnel tests and NACA, the forerunner of NASA, advice.

A high-level U.S. delegation visited the plant in late 1955 to examine the airplane program in detail. During their visit, the RCAF emphasized the *Arrow's* importance to the defense of both the U.S. and Canada and suggested some U.S. financial support one way or another. The delegation considered the design technically sound and agreed with Avro's performance estimates. They thought that achieving a 2g turn at Mach 1.5 was a bit ambitious, though a 1.2g maneuverability would meet their requirements, and even 21st-century fighters have difficulty beyond that.

Subsequently, on November 9, 1955, the U.S. Secretary of the Air Force, Donald Quarles, sent a letter to Canadian Minister of National Defence Ralph Campney saying that as a result of their evaluation, they recommended that the CF-105 *Arrow* and their Orenda *Iroquois* engine developments and production should proceed as planned. This U.S. interest was further confirmed in early 1957 by USAF General Boyd, Chairman of the Advanced Interceptor Committee when he asked to be kept apprised of all developments. Later that year, Dr. Courtland Perkins, USAF Chief Scientist, told Jim Floyd, Avro Vice President Engineering, that the USAF had awarded a contract for an airplane similar to the *Arrow,* the F-108. But support for it was diminishing because it was heavier and costlier, so the USAF would be interested in buying the *Arrow* if the F-108 was cancelled. However, the U.S. Secretary of the Air Force said in January 1958 that the F-108 was being cancelled and that the *Arrow* would NOT be needed!

The short-range *Bomarc* ground-to-air missile system was now being seriously considered, and the general view was that it should be deployed along with long-range interceptors. Still, was a growing concern that the Canadian budget could not afford both.

The F-108 fighter was cancelled because an influential group in the Pentagon concluded that its role could be accomplished quicker and at far lower cost by the *Bomarc* and because it would have been useless against the growing threat of an Intercontinental Ballistic Missile (ICBM).

The North American Air Defense Command (NORAD) was established in August 1957, and the following May, the Canadian and United States of America governments agreed to cooperate in their joint defense. The first order of business was setting up the *Bomarc* system and its associated Semi-Automatic Ground Environment (SAGE). The latter was a complete surveillance and weapons control system. It involved a series of installations across Canada roughly along a line parallel to the border and about 200 to 400 miles north of that border.

Canada committed itself to pay for and supporting a portion of SAGE, with the full knowledge that they couldn't afford it! The U.S. needed it to protect its ICBM bases and northern cities, but Canada could have protected its vital areas with the *Arrow* alone, merely using SAGE to vector the aircraft to its target. In fact, the *Arrow's* radar enabled it to operate beyond the SAGE line.

However, the die was cast. Canada's Minister of National Defence, Roy Pearkes, who'd replaced Campney, asked his U.S. counterpart what the U.S. would do if Canada refused to install the *Bomarc*. He was told bluntly that the U.S. would merely move those bases south of the border. In that case, the *Bomarc's* nuclear warhead would deposit its radioactive remains over Canada's highly-populated areas if an intercept was made there. This concerned Mr. Pearkes a great deal! The basis for the continental defense was that the *Bomarc* and Canada was bamboozled into allowing the Americans to install their SAGE system across Canada. President Eisenhower and Prime Minister Diefenbaker had a three-day meeting but it seems that Dief was blinded by his admiration for Ike, and he lost the opportunity to work out a deal that would have had the *Arrow* playing an important role in their overall defense plan.

Defence Minister Pearkes fluctuated wildly in 1958. First, in discussion with U.S. Secretary of State John Foster Dulles, he told Dulles that Canada had a supersonic fighter that both countries thought was important in their joint defense and asked for some cost-sharing since the U.S. was asking Canada to help pay for SAGE. He noted that the *Arrow* did not need SAGE, but Canada added that system to facilitate *Bomarcs* and some U.S. aircraft. Then later, Pearkes told the Canadian Cabinet that he'd agreed to recommend two *Bomarc* sites near Ottawa and North Bay. HE'D RECOMMEND CANCELLATION OF THE *Arrow* BECAUSE *Bomarc* WOULD COST A LOT LESS, but would be as effective as the *Arrow*. Also, the U.S. F-106C would be able to do everything the *Arrow* could do at a lower cost and in far less time.

We don't know where Pearkes got those stupid ideas. First, SAGE was not needed for the *Arrow's* mission. Second, the *Bomarc* would NOT be as effective as the *Arrow*. Its range was a lot less, and there was always the possibility that it could shoot down a commercial airplane by mistake. And lastly, the F- 106C did NOT have performance comparable to that of the *Arrow*.

All of the above illustrates the shameful non-technical demise of a world- class airplane, brought on mostly by politicians who were incapable and unwilling to find an American/Canadian solution for their joint defense. The Americans began by strongly supporting the project, and they obtained a great deal of technical knowledge from it. Their support dwindled with the realization that U.S. defense industries would cause an upheaval in Washington if the USAF bought a Canadian fighter. The Canadians had bad faith in their own capabilities, listened to much erroneous advice from so-called experts, and completely gave up trying to find an acceptable solution for the Americans. At the very least, the *Arrow* should have been funded until the *Bomarc* was tested to evaluate its capability. That missile was subsequently cancelled, leaving a dangerous hole in the defense system that was fortunately not exploited by the Soviet Union.

After thinking about this for some time, I believe a solution could have been reached. Canada would agree to pay for its *Arrows* and allow but not contribute to the SAGE installation provided that *Arrows* could connect with it. Canada would also not pay for *Bomarc* sites on Canadian soil; if that meant them having to be located south of the border, then so be it. The Canadian fighters would take care of any intruding Soviet aircraft.

On February 20, 1959, despite assertions from the RCAF chiefs, such as Air Marshall Slemon, that an interceptor would be needed for some time to come (and now, more than 50 years later he's been proved right), Prime Minister John Diefenbaker cancelled the *Arrow* aircraft and the *Iroquois* engine programs. The reason he gave was that the threat had changed. Manned bombers would be replaced by ballistic missiles, against which the *Arrow* was useless. (Note: He didn't mention that the *Bomarc* was also ineffective against these). He then quoted the price tag for the *Arrow,* but these prices were inflated by about 30 percent beyond their correct value, probably because they were prepared by someone who had not examined the latest numbers, or for reasons unknown, had purposely twisted the numbers to suit Diefenbaker's argument.

With their funds eliminated, Avro decided immediately to close their aircraft end engine plants. Workers were told that the plants would close that Friday night, but they could return on the Monday to retrieve their belongings. Over 14,000 Avro employees and more than 10,000 subcontractor employees were laid off overnight! There was also the "domino" effect among local merchants when restaurants and gas stations suddenly lost their customers, and there was a glut of houses that suddenly went for sale with consequent drops in prices for those homes. Workers began scrambling for work elsewhere, and the U.S. and the U.K. became prime beneficiaries of many engineers. Jim Floyd became the Chief Engineer of a Hawker Siddeley advanced project group in the U.K., and Chief Aerodynamicist John Morris joined him. Jim Chamberlin joined NASA to take charge of engineering on the Gemini space program, bringing a team of 25 men he had handpicked Chief Engineer Bob Lindley went to McDonnell Aircraft to be their number one guy on the Gemini. Several other top guys found good jobs at Boeing, North American, and Lockheed.

Finally, to make the whole ridiculous mess worse, the Canadian government realized that they still needed new fighters within four months of the cancellation. They issued a contract for 214 Lockheed F-104s to be built by Canadair. Later, they bought other aircraft from the U.S. to supplement and replace the F-104. The *Arrow* and its subsequent development would have obviated the need for these aircraft. Rarely, if ever, has there been such a lack of leadership, negotiating skill, national pride, and broad perspective of overall defense on the part of any government, and the associated awarding of a highly complex contract and subsequent procurement activity.

Avro Canada CF-105 Arrow

CHAPTER 10
HERC. AT 60

The Lockheed C-130 *Hercules* was the world's first real airlifter, and today we find it operating in more than 70 countries. Since its first flight in 1954, we have seen the C-141 *Starlifter* provide additional payload and range, the C-5 *Galaxy* accommodate the army's largest tanks and six Greyhound buses if such a need ever existed; the C-17 *Globemaster* do some of the C-5's work at a lower cost; the Airbus A400M fill a gap that Airbus perceived between the C-130 and the C-17, and the Alenia C-27, a sort of mini-Hercules. These were designed to fill a niche in airlift, but after surveilling the entire field, the *Hercules* stands out as the best suitable for general airlift roles, even after 60 years.

Why is it the most successful? First, credit must be given to those in the USAF who prepared the requirements in 1951 and 1952. They were thoughtful enough to ask for the ability to quickly drive vehicles in and out of the cargo compartment and airdrop huge pallets of cargo. They allowed 64 paratroops to be dropped from it; to operate from short, rough, unprepared strips, and to have enough speed plus payload/range to operate to all parts of the world. Specifically, four typical missions were defined. One example is it had to carry a 25,000-pound payload for 1,100 miles, fly over the combat area at high speed, low altitude, and return without refueling. The second clue to its success was the approach taken by the design team, led by

Willis Hawkins. They abandoned all preconceived notions of cargo-plane design. Instead, they considered a rugged structure with a large cargo compartment accessible by an aft ramp from which trucks could drive straight in, instead of the old side- fuselage cargo door that needed ground equipment. They visualized a simple landing gear with large tires to overcome rough and soft surfaces. For the power, they decided to use the economical but powerful Allison T56 turboprop engines which were relatively new at that time. The third reason for its success is its suitability for continuous development, and Lockheed taking advantage of that over the years by constantly upgrading it and adapting it for a wide assortment of missions such as airlift, cargo dropping, gunship, medical evacuation, tanker, coast guard, and even commercial passenger transport.

Boeing, Douglas, Fairchild, and Lockheed competed for the contract, and Lockheed was awarded it on July 2, 1951. Two aircraft were to be built, and on August 2, 1954, the first YC-130 took off from Edwards Air Force Base, California.

The first aircraft were noteworthy in several respects. They had integrally-machined cargo floors and wing skins, which were unusual at that time, and a new high-strength aluminum alloy. The aft entry had a lower ramp and a hinged upper door to provide an opening as large as the cabin height, and the cargo floor permitted truck-bed height loading. Another unusual feature was cabin pressurization needed for its medical evacuation role, carrying up to 74 stretchers. The extra-large tail fin enhanced stability and control at low speeds. The flight crews welcomed extra-large windows to help in observation under challenging environments, and in later versions, such as coast guard, they were particularly valuable in search and rescue.

All C-130 activities were transferred from Burbank to Lockheed-Georgia in 1952, In April 1955, the first production model flew at Marietta. Test pilot Leo Sullivan said, "Takeoff power was fantastic. We knew that here was one of those happy circumstances of shape, structure, and propulsion where everything fits together just right;

110,000 pounds gross weight, 3,750 horsepower per engine, with instant response to power levers, and a 3-g load factor for rapid military maneuvers." The enthusiastic engineering team then began a years-long succession of improvements to the airplane and its evolution to many unforeseen roles. The first model, C-130A, had a gross weight of 108,000 pounds, but this was soon increased to 124,000 pounds, and the original T-56-A-1 engines were replaced with more powerful T-56-A-9s.

The first four C-130As were delivered to the USAF's Tactical Air Command (TAC). Several variants were introduced quickly after that, one of which added skis to the landing gear (C-130D), ostensibly to move supplies to the Distant Early Warning (DEW) Line, the chain of radar stations along the northern edge of Canada, way north of the Arctic Circle. Once it had shown everyone how it could do that kind of work, it was then used to move oil well equipment in Alaska and make several missions to the U.S. Byrd and Pole Station Antarctica. These skis measured 20.5 feet long and 5.5 feet wide for the main landing gear, and 10.3 feet long by 5.5 feet wide for the nose gear. In total they weighed 5,600 pounds and were neatly retractable, so that range was only reduced by 9 percent. The USAF bought 13 of these aircraft, but one was later transferred to the U.S. Navy.

The first mission to the South Pole area was on January 23, 1960, and it was summer down there. Seven of the ski-equipped C-130s landed smoothly one after another at Williams Field skiway and plowed through snow to a stop. Having paved the way, within a couple of days, a series of 58 shuttle runs began from Christchurch, New Zealand, to supply bases near the Pole. Some 400 tons of desperately-needed supplies were taken in within two weeks in temperatures that ranged from -40°F to -8°F. Later, the U.S. Navy obtained its own ski *Hercules*, the four red- tailed C-130BL, subsequently designated LC-130. They formed what was later known as the "Penguin Airline." The stories of their many flights are legendary, often disgorging cargo in blizzard conditions.

Another C-130 variant was the *Gunship* (AC-10A and JC-130A) that was precipitated by the Vietnam War. As I remember it, the U.S. wanted some way to attack the North Vietnamese forces that constantly drove south on the Ho Chi Minh Trail at night. Some bright guy got the idea of installing a variety of weapons on a C-130 with infrared target acquisition to detect the target. The result was quite remarkable. With two 20mm multi-barrel rotary cannons and two 40mm cannons, they would circle the target and wreak havoc with the convoys of trucks below. One pilot said that, in his opinion, the Gunship was more effective than fighters or bombers. They had more firepower and could stay up there far longer than any fighter, which more or less whizzed by a high speed. On the other hand, the Gunship circled at a lower altitude, with up to seven guys looking for targets.

The Drone Launcher was another variant of the "A" model, designated originally as the GC-130A and later as the DC-130A. It carried drones such as the Ryan *Firebee* on pylons beneath the wings. These were launched and guided (microwave) for various missions such as reconnaissance, electronic intelligence, strike, and targets. That same model was also used to track missiles and space vehicles.

By 1951 the *Hercules* had proved itself, and engineers began improvement studies culminating in the "B" model. The engines were replaced by T56-A-7s developing 4,050 eshp driving four-bladed Hamilton-Standard 13.5 feet propellers having lower noise and less vibration. Gross weight went up to 135,000 pounds, and range went up by 250 nm. This model made its first flight at Marietta on December 10, 1958. Like its predecessor, the airplane was adapted to fill several roles, one of which was a tanker version for the U.S. Marine Corps. Another was for the "Hurricane Hunters" in the Atlantic area and "Typhoon Chasers" over the Pacific, with a similarly equipped version for the National Oceanic and Atmospheric Administration's studies of tropical weather.

The USAF awarded Lockheed a contract for a C-130 STOL version of the C-130B in 1958, and this generally became known as the BLC

(Boundary Layer Control) version. The U.S. Army wanted to carry a 20,000-pound payload for 1,000 nm, with a takeoff and landing roll of 500 feet on an unprepared surface. The airplane used high pressure blowing over the flaps to increase wing lift after changing the flaps from their original Fowler type to a simple-hinge type. The air was provided by two Allison T56 engines in underwing pods. The wing lift was increased substantially, decreasing stall speed so much that extra control was needed. To do this, BLC was also applied to the ailerons, rudder, and elevators. Overall, this proved to be successful, but the contract was cancelled as part of the defense department's cost-cutting. Using its own money, Lockheed flight tested the airplane in early 1960. In one case, at 105,000 pounds landing weight, the stall speed was way down to 50 mph, and its landing roll was only 450 feet.

At this point, the USAF Military Air Transport Service (MATS) enters the picture. They had been operating antiquated aircraft for some time and were consequently a prime focus for Lockheed's marketers when they came up with what-they-called an "advanced C-130B," which later became the C-130E. MATS wanted an airplane that could carry a reasonable payload over the Atlantic or Pacific. The new airplane did that! By adding two 1,300-gallon (U.S.) external fuel tanks below the wings and beefing up the structure to match a 155,000 pounds gross weight, they were able to carry 45,000-pound payload for 2,200 nm, or 20,000 pounds for 3,000 nm. With numbers like this, MATS bought it, and on August 25, 1961, the "E" made its first flight.

The C-130E had the same power plant as the "B," but its longer range and higher payload made it a best-seller both in the U.S. and many countries around the world. Its operations in Vietnam showed the world how useful it would be in various roles. For example, it showed how large cargo pallets could be dropped by parachutes and huge loads with high accuracy from altitudes of about 10 feet.

Rescue and recovery were vital—rescuing downed airmen and recovering cassettes/film packs from space vehicles. In the rescue role, an HC-130H had a Fulton Recovery System mounted on the aircraf"s nose. This comprised two aluminum alloy forks extended forward from their stowed position to form a "V" with a 60-degree angle. A harness and an attached balloon would be dropped to the person being evacuated. After wearing the harness, filling the balloon and deploying it with its nylon tether, the C-130 would fly over, catch the tether at a height of less than 500 feet, and pull in the airman through the aft cargo compartment door.

All of which brings us to the C-130H—the primary workhorse for many years and the most-produced version of the C-130. It first flew in 1964, and relative to the "A" model, it had 26% more payload, 11% more speed, 52% more range, and 17% less takeoff distance. Variants included the basic transport, search/rescue, tanker, weather reconnaissance, and an advanced gunship, known as the *Spectre*. This gunship had two 20 mm rotary cannons, two 7.6 mm machine guns, and a 105 mm howitzer, plus a laser target designator. The structure of the airplane was strengthened, and engines were updated. New materials were used for corrosion protection, and flight controls were modified to increase reliability and provide further backup.

In addition to the C-130H purchases by the USAF, U.S. Navy, U.S. Marine Corps, and Coast Guard, it was also purchased by 49 foreign governments, including the Royal Air Force (RAF). The RAF order was for 66 aircraft, but they were significantly modified, so they were designated as C-130K. They had a cargo floor specially adapted to British equipment, and British avionics were installed. Since the RAF order was so large, a British company, Marshalls of Cambridge, was authorized to conduct major maintenance and modifications. A prime example of the latter was the introduction of aerial refueling, a necessity for their conflict in the Falklands Islands, thousands of miles south of the U.K. Thirty RAF C-130s were modified to the "Dash-30" length—a 15-feet stretch in the fuselage. This allowed

128 infantry troops to be carried, compared to 92 in the standard version. Seven cargo pallets could be carried versus five originally.

Like its U.S. counterparts, the RAF aircraft were used in a wide variety of rescue and humanitarian missions. In one case, during the disturbances in Cyprus, they evacuated 7,707 people, and in one flight, they accommodated 139 passengers.

While all of these modifications were underway in the 1960s, engineers began exploring commercial variants of the aircraft, designated the L-100. They employed various degrees of fuselage stretch to suit particular applications, and some were sold for cargo hauling or moving heavy equipment to mining or oil activities in remote areas. There was also the L-400 mini-*Hercules*, powered by two engines. There didn't seem to be a good market for this version, so it was dropped.

There were ten customers for the L-100-10, twelve for the -20, and15 for the -30 *Super Hercules*, having cargo floor lengths of 41 feet, 49 feet, and 56 feet, respectively. They began flying in April 1964.

It was logical also for the preliminary design department to study many other variants, including replacing the turboprops with turbofan engines. They also studied a version called C-130VLS for increased volume, loadability, and speed; a trijet tanker called the KCX-130; a version that had a larger fuselage cross-section and STOL capability (the WBS), and there was even an amphibian, called the *Hercules HOW* (Hercules On Water).

There were many studies of design enhancements on the military side that could be incorporated into the basic design to make it a competitive airlifter for the 21st century. It was apparent, for example, that the flight station needed to be updated, and short-field performance improvements were desirable. The airplane that emerged from this was the Assault C-130, and it had everything conceivable for a tactical airlifter. Naturally, the various armed forces, at the operational level, were enthusiastic about it, but then the budget-conscious people decided to examine those improvements that were the ones worth paying for.

Eventually, the airplane that resulted from this was the C-130J— the latest Herc incarnation.

The RAF was the launch customer, and by mid-1995, the first C-130J rolled out. It had an advanced flight station, more powerful engines, and advanced six-bladed propellers. The new Rolls-Royce AE2100D3 engines provide 29% more thrust, and fuel efficiency is increased by 15%. This results in more speed and a 40% range increase relative to the "E" model. In the flight station, there are four LCD, and both pilots are provided with holographic heads-up displays. Onboard computers monitor the various systems, and this, along with greater overall reliability, results in substantial decreases in maintenance manpower requirements. Fuselage length is 15 feet longer than the original; i.e., the "Dash 30" length, referred to previously, results in superior load- carrying capability.

Currently, there are the basic transport model, the HC-130J Coast Guard, the WC-130J "Hurricane Hunter," the EC-130J psychological warfare model, and the KC-130K aerial refueling variant. Since this is the current model, they are still being sold worldwide, with over 300 delivered or on order from 16 countries.

So as the legendary C-130 passes its 60th birthday, it is notable that more than 2,500 have been produced for 63 nations. More than 70 variants have been built, and it is the longest, continuous military production run in history. One of the top three longest continuous aircraft production lines of any type. It has operated from the North Pole to the South Pole and around the equator, and it has operated from the highest Himalayan airstrips and the deck of an aircraft carrier.

C-130 Gunship

C-130 at South Pole

C-130J

CHAPTER 11

BOEING'S BIG ONE & THE COST COMFORT DEBATE

The advent of the Boeing 747 brought the debate that is still ongoing. Airliners had been growing over the years, but the 747 seemed to be one huge leap ahead in this regard. As one entered the cabin, it seemed that rows and rows of seats presented themselves for the masses of travelers who were now about to explore the world. What a change from the 1950s! No longer was air travel the refined way to cross the continent or the ocean, dressed appropriately, and being wined and dined with grace and expertise by attendants. The picture changed. Fares plummeted to levels that were now affordable by average middle-class families, so they now hopped into a plane to tour Paris in the spring, dressed in golf shirts, shorts, and sneakers. All of this is probably beneficial to the world at large, but it has caused a great deal of controversy worthy of some discussion. But first, a look at the genesis of the 747 is interesting.

The 747 story began in October 1961 when the USAF issued a requirement for a C-133 replacement, and the Pre-Concept Formulation Phase began. The project was then called the CX-4 and later the CX- HLS (Heavy Logistic System). After two years of studies by major aircraft companies and the USAF, a Specific Operational Requirement (SOR) was issued. Some of its basics were as follows:

- Carry a payload of 100,000 to 300,000 pounds for 4,000 nm.
- Cruise at 440 KTAS at 30,000 feet altitude.
- Takeoff distance 8,000 feet at sea level at maximum gross weight.
- Landing distance 4,000 feet with 100,000 pounds payload and fuel reserves for 4,000 nm
- Capable of landing on support area airfields (i.e., "unprepared" surfaces).
- Cargo compartment 100 to 110 feet long, 16 to 17.5 feet wide, 8.5 feet high, able to accommodate two rows of palletized cargo.
- Straight-through loading (nose and tail).

Using these requirements, a series of configuration studies was conducted, and this included the consideration of two new turbofan engines, one by General Electric and one by Pratt & Whitney. Since the U.S. Department of Defense was headed by Secretary of Defense Robert McNamara, a man known for his detailed systems analysis, it was necessary for the companies to conduct extensive trade studies so that every major decision was backed up with reams of paperwork. While this is a fascinating example of thoroughness, it was often exasperating to engineers who, in the past, had relied on experience and instinct to decide between various design approaches, but now had to spend many, many hours proving it.

For example, would it be better to have 36 relatively small wheels or just four very large wheels on the main landing gear or somewhere in between? The weights of the two designs would be compared, along with the structure to house them, the relative maintenance and reliability involved, the safety aspects, manufacturing costs, and the relative performances on the landing field being considered. Increased weight causes more fuel to be burned, more elaborate structure could be heavier, and if increased pod housing is used, this causes more drag that impairs performance and increases fuel usage. The comparison is eventually consolidated by evaluating lifetime costs. Almost all aspects can ultimately be related to cost—the fuel used, maintenance manhours, parts replacement and so on. When these are all considered the conclusions can sometimes be quite remarkable, and some of the "old-timers" were quite perplexed when their intuition proved incorrect.

The HLS was a huge increase in size, weight, and operational capability, so it was inevitable that the many tradeoffs and configuration studies would be lengthy. It took until April 1965 for the proposals to be completed. The USAF selected the winner in October 1965. It was Lockheed, and it became the C-5 *Galaxy*.

Not to be outdone, Boeing paid attention to overtures from Juan Trippe, Pan Am's President, who'd been advocating an airliner of HLS-size for some time. Since Pan Am was, at that time, the world's biggest airline with routes circling the globe, Trippe's ideas were of some consequence, and Boeing realized that their HLS design could be born again – in civilian clothes. Think of the ramifications of this. Boeing lost the C-5 contract but, as a result, produced the 747 that ultimately would be of far more significant financial benefit to the company.

Boeing's design still had the elements of a cargo transport, such as an easily-convertible cargo nose in the front. The idea is that they may still find a market for it as a cargo carrier if and when its sales as an airliner were not up to par.

In April 1966, mere six months after Lockheed had won the C-5 contract, Pan Am ordered twenty-five 747s, but other airlines didn't respond immediately. American Airlines thought the 747 was too big and got together with Eastern Airlines and TWA to develop their own requirements. This resulted in the tri-jets from Douglas and Lockheed, the DC-10 and L-1011, operational within a year of the 747's debut. These three airliner types satisfied the world's demand for large airliners in the 1970s and 1980s, and it is interesting to note that the 747 outlived all others. It was the first wide-body, and with the DC-10 and L-1011 out of production, it is still going strong!

While redesigning the fuselage for airline use, Boeing kept the flight station perched above the cargo deck but decided to not have an upper passenger deck, partly because of the difficulty in meeting FAA emergency egress requirements. So, the ubiquitous hump was there from the beginning. At first, the area behind the flight crew was used as a small lounge, but first-class, business, and elite passengers were accommodated there later, and the hump extended. In January 1970, the 747 entered service with Pan Am, and it has proved to be a very popular airplane with passengers and airlines. It has had an excellent safety record, and more than 1500 have been built, with more on order. It cruises at Mach 0.85 (570 mph), and the −8 version can carry 467 passengers in a 3-class configuration for more than 8,000 nm. However, it is noteworthy that on one occasion in 1991, it carried 1,081 passengers.

For those of us who'd been flying transoceanic routes on Boeing 707s and Douglas DC-8s, it was quite a surprise when we first entered a 747 cabin. Instead of a single aisle running down the center of a tube, we now saw two aisles in a double-wide cabin, with rows and rows of seats on all sides. The space seemed even more pronounced in first class, and the airline was delighted with the low cost per passenger mile.

Eventually, the cost of comfort began to enter the picture. The speed or comfort factor must be recognized; i.e., if you're crossing the Atlantic at a supersonic speed, your time aloft is so short that you should be willing to accept somewhat cramped conditions. However, if you're flying between Asia and California at subsonic speed, it takes many hours, and more comfort, maybe at a higher cost, should be considered. The Anglo-French *Concorde* is an example. It zipped across the Atlantic at 1,350 mph, more than twice the 747's speed, and its passengers were quite happy to be sitting four abreast in a relatively long thin tube when completing their trip so quickly.

But passengers began looking for alternatives, and Business Class was offered—a great idea. It provided affordable fares with comfortable seats. The airlines are quite cognizant of the public's discontent with knees jammed up against seat backs and reclining seats impacting their use of trays for laptop use, and they are responding with a policy of "you get what you pay for." The result is probably the closest possible approach to the overall perception of airline comfort.

Taking one typical airline as an example, Delta Airlines' 747 service has 40 Business Elite seats at 80 inches pitch, 42 Economy Comfort, sometimes generally referred to as "premium economy," seats at 35 inches pitch and 286 Economy seats at 32 inches pitch. The Business Elite seats are in cubicles, providing privacy, less noise, flat-bed seats, bedding, attractive meals, and several other amenities at a higher price. If you're on an expense account, if you're wealthy enough, or if it is a special occasion that you're willing to splurge for, then this is a viable option. Economy Comfort may suffice if you can't afford this but want more legroom, more seat-reclining capability, and preferred cabin location at the front of the economy section. We haven't found an airline that offers the luxury of pre-war flying boats, but we wonder if such luxury with the attendant high fares is warranted in today's high-speed world.

Boeing 747

BAC/Aerospatiale Concorde

CHAPTER 12
WHAT'S UNDERNEATH

Sometimes it's called the landing gear or the undercarriage. A USAF specification even called it the Alighting Gear! Whatever you want to call it, the general public pays little attention to it until that rare occasion when something goes wrong, often with disastrous consequences. Perhaps the reason for its poor publicity is because of the technical jargon used relating to brakes, tires, hydraulics, bogies, shock absorption, and so on compared to things people understand such as seats, windows, air conditioning, galleys, and maybe even the flaps and ailerons that they can see moving. Since this book is intended to provide interesting non-technical discussions to those who have a general interest in aviation, this chapter explores the peculiar history of the landing gear and some of the challenges and solutions involved, and an attempt is made to avoid technicalities.

It was noted earlier that Sir George Cayley invented he bicycle-type wheel for use on airplanes in 1808 and developed a successful glider. About 50 years later, the Wright Brothers made their first flight. Between 1906 and 1910, the early pioneers installed crude landing gears on their aircraft, and World War I resulted in a major step forward when "proper" landing gears appeared—wheels mounted on V-shaped struts from the underside of the cockpit and a tail skid. Typical examples were de Havilland's B. E.2, Bristol F2B, SE 5, Nieuport 12, Sopwith *Camel*, Spad VII, and Fokker DVII. They didn't have brakes, but they did have some shock absorption, typically bungee cords wrapped around the axle and the support structure, and in the case of the *Camel,* it allowed four inches of upward wheel movement.

The tires used on the first airplanes were similar to those used on bicycles, but they have evolved to enormous sizes with very advanced arrangements of treads. The largest I found was six feet in diameter! Generally, though the airplane's weight is better supported by having a larger number of smaller tires since it spreads out the load so that the runway is not damaged, the space to house the retracted gear is more compact, and there is a safety issue. If one tire bursts on a multi-wheel gear, it is far easier to accommodate than a single tire bursting.

As airplanes became bigger and heavier, their takeoff and landing speeds increased to provide enough lift, necessitating brakes. In the late 1920s, the first brakes were developed by Palmer in the U.K. and then by Dunlop. BF Goodrich developed them in the U.S.

This same size and weight increase also required better shock absorption, and to grasp its significance, imagine what would happen if a two-ton pickup truck was dropped, from, say, ten feet and expand to a 400- ton airliner landing. Even taxiing can cause passengers to feel up to nearly 2g's on current airplanes with sophisticated shock absorbers. The conventional air/oil shock absorbers began in the mid-1920s, with well-known companies Bendix, Dowty, and Messier leading the way. Since then, the basic essentials haven't changed much but there have been advances inside the shock strut metering the oil to achieve high efficiency, and careful selection of materials for high efficiency for the least weight. It should also be noted that rubber blocks have been used to obtain satisfactory results in several applications. The Ford Trimotor of 1932 used a stack of these blocks to provide adequate damping, and later in 1942, the de Havilland *Mosquito* used them to do the job at low cost and minimum use of skilled machinists. On some light aircraft such as the Cessna 172, the landing gear strut was essentially a leaf spring.

The spring had the wheel at one end and was attached to the fuselage at the other end—the simplest and least expensive shock absorber.

I designed a leaf-spring gear for a bush 'plane back in the 1950s and found, that although it is simple, the devil is in the details. My first attempt had the wheel moving the correct distance when the load was applied, but the spring wasn't strong enough, so a delicate balance had to be found like changing the spring cross-section that had just enough strength and the appropriate "springiness."

Airplane performance increases lead to the desirability of retracting the gear to minimize drag, but it was also realized that this would increase weight and cost. To "house" the gear a hole must be made in the wing, fuselage or engine nacelle, so all added complexity must be evaluated. Increased weight means more fuel used and payload cut, and the provision of space to house the retracted gear disturbs the idealized structure, adding more weight and cost. Consequently, there is usually little to be gained by retracting the gear on a low-speed light airplane apart from aesthetics.

Among the first aircraft to have retractable gears was the Bristol *Jupiter* racing airplane in the late 1920s, the Lockheed *Altair* of 1930, the Boeing V1B-9 bomber of 1931, the 1932 Grumman FF-1 fighter, which tucked up its wheels into the fuselage side, the Douglas DC-2 and the Boeing 247-D. Apart from the FF-1, the systems used were simple; i.e., none had the ingenious systems we see today that skew, rotate, and fold the various parts to fit into an ultra-small cavity.

It is worth noting that there was an in-between phase in which landing gears were not retracted but were covered by streamlined spats, and there are many of these. Wiley Post's Lockheed *Vega* and the Gee Bee *Super-Sportster* (1932) are typical examples.

From the design standpoint, the "armchair engineer" should consider some of the complexities of a retracting gear. First is the structure that

151

supports the gear. All of the braces and struts that must be collapsed appropriately and locked to guarantee that the gear is locked down or up. When the pilot selects "gear up," the locks must open, the wheels should stop spinning, and the bogie that holds the wheels must be appropriately oriented to fit into the cavity. The actuators must then raise the entire gear into its cavity, uplocks must hold it in place, and the doors should close all within a few seconds. Imagine the engineer's tension as he observes all of this happening to his landing gear for the first time.

Before that first retraction, the engineer had used mathematics and precise geometric layout to ensure that the gear would fit into its cavity with just enough clearance from its surrounding structure. For the first test the gear is retracted very slowly, stopping periodically to check all of the linkages, and eventually, it pops into the cavity. If all goes well, it has satisfactory clearances, and none of the struts or braces have poked through the adjacent skin or fuel tanks. If there is any interference, then it's back to the drawing board, and changes at this stage could be costly and time consuming.

There is another type of landing gear that deserves mention, at least for historical purposes, and that is the air cushion. Essentially it is a large inflated bag beneath the fuselage, and most of the work on it was done by Bell Aircraft. In some ways, the concept seems obvious, and it has been used successfully on boats. Bell developed a small airplane, the LA-4, that demonstrated some noteworthy achievements when operating over water and marshland. They followed that with a larger vehicle, an adaptation of the DH Canada *Buffalo*, but the weight, complexity, drag and stowage concerns were too much for it to be developed any further.

With airplane weights increasing, it became necessary to recognize runway strength. The main landing gear has only two struts on most airplanes, one left and one right. On both the Boeing 747 and the Lockheed C-5, there are double that number, and on the C-5, there are six wheels on each strut, plus four wheels on the nose gear. The huge Airbus A380 and the Russian TU-144 have similar arrangements, and all of this is necessary in order to spread out the load on the runways and taxiways. Otherwise, repeated landings and takeoffs would damage the surface, and consequent repairs would not only be expensive but would severely impact use of that airfield by other aircraft.

The C-5 *Galaxy* was a particularly interesting case because it was also required to operate from bare soil fields. At a weight of nearly 300 tons, it was required to make at least 130 landings and takeoffs on bare soil before the surface needed repair. That soil would have the consistency, roughly, of a dry football field, i.e., not a highly compacted hard surface, but better than a muddy field. Hence the need for an unusual tricycle arrangement of six wheels on each strut.

Not only did it have to have to operate from such fields, but it had to kneel the airplane for efficient loading and unloading. Ballscrews on each strut lowered the fuselage by 34 inches after landing until the belly was almost touching the ground, with the cargo floor down to truck bed height, so no ground equipment was needed to move cargo to trucks or other loaders. The C-5 also had some unique features such as onboard tire inflation or deflation and the ability to rotate each strut 20 degrees left and right to facilitate crosswind landings without using the awkward sideslip maneuver usually employed. It was also required to land on surfaces that had six-inch bumps, which had some effect on tire selection and the use of a specially-designed shock absorber.

The four-strut arrangement also enables one leg to be raised (via ballscrews) to be repaired or serviced while the airplane is well supported by the remaining legs. There is a safety dividend here, too, in that, to some extent, a landing can be made with one or more gears inoperative.

Finally, in the CX-HLS competition, Lockheed did an extensive tradeoff study to determine the best landing gear arrangement. Six hundred sixty landing gear variants were considered for each of three aircraft weights. They studied wing- mounted, underfloor-mounted, side-fuselage designs, and six different bogie arrangements with an assortment of tire sizes. Although the bogie with three pairs of wheels, one pair behind another, appeared instinctively to be the best. The twin-tricycle turned out to be the winner, partly because the ruts in the soil were spread out more.

So, in conclusion, we can see the tremendous metamorphosis from the simple World War I two wheels and a tail skid to the huge 28-wheel arrangement of today on an airplane that can disgorge two main battle tanks on a bare soil field.

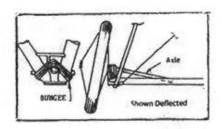

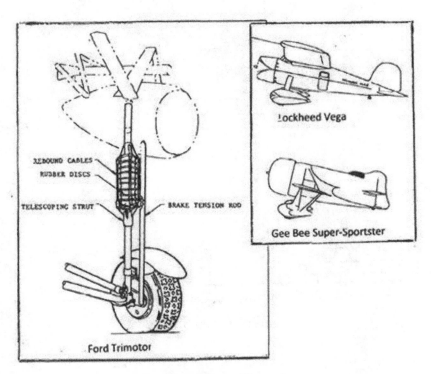

Landing Gears Sheet 1

155

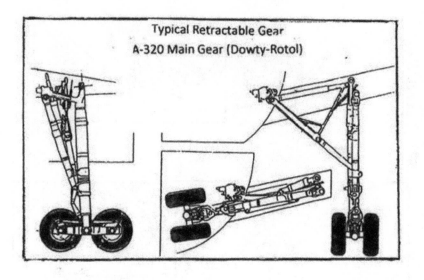

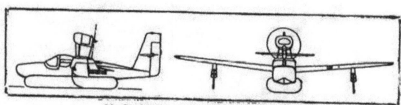

Air Cushion Landing System (Bell LA-4)

Landing Gears Sheet 2

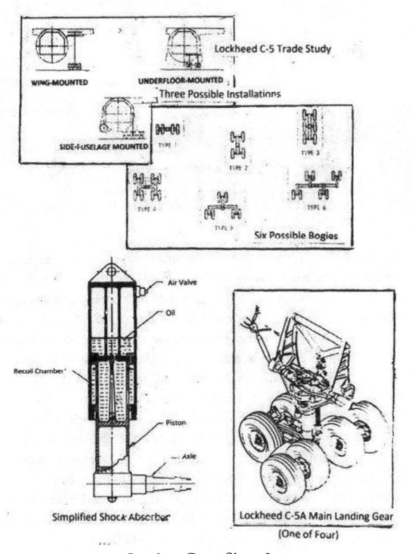

Landing Gears Sheet 3

CHAPTER 13

WHERE ARE THEY NOW... AVRO, BRISTOL, CONVAIR...

In World War II, more than 70 American and British companies were designing and building airplanes. The German aircraft industry was just as robust but was concentrated on fewer companies, as were those in the USSR and Japan. When fighting stopped, the funds did too, and there was a huge supply of surplus planes. Some smaller companies just fizzled out, and larger companies shrunk to adjust to the new reality of a smaller market. With this in mind, many of us wonder what happened to some of them. One resource is the book "The Complete Illustrated Encyclopedia of the World's Aircraft" by David Mondey. However, it includes a very extensive summary of ALL companies, large and small, very old and very new, many of little consequence to this particular discussion. In many cases, it does not say what happened to the company. It must also be recognized that this book was published in 1978 and does not include changes since that date. So some diligent research is in order.

In the ensuing pages, some companies have been omitted because their lives were too short or they were not among the major players in the overall scene. Among these are Boulton Paul, Bellanca, Chilton, and General Aircraft.

The British aircraft industry saw the greatest upheaval, and in retrospect, this was inevitable. Their market is far too small to accommodate the large number of World War II companies. The large U.S. population provides a proportionately large budget to buy large numbers of airplanes for their armed forces. With its long distances, it encourages the use of commercial air travel. These factors do not apply in the U.K. Its much smaller budget can't afford the high front-end costs of designing and developing many types of advanced aircraft. With its domestic airline limited by the competition of efficient high-speed rail, it must generally concentrate on international flights. For this, they generally resort to the more cost-effective approach of buying U.S. aircraft or taking part in Anglo-French cooperative projects.

One of the RAF's best long-range bombers in the early days of World War II was the *Whitley*, made by Armstrong Whitworth, a company dating back to 1914. After the war, it developed a two-seat night fighter and a turboprop airliner, the *Argosy*, after which it became part of the Hawker Siddeley Group and then BAE Systems.

Hawker Siddeley/BAE Systems also incorporated Hawker, famous for its *Hurricane, Hunter* and *Harrier,* the de Havilland, which well-known for its *Albatross, Mosquito* fighter, bomber, etc., and the *Comet* jet airliner, Avro with its *Lancaster* and *Vulcan* bombers, and Blackburn that made the *Skua* and *Buccaneer* for the Royal Navy. A note for the history buffs—Sopwith, famous for its World War I *Camel, Snipe,* and *Pup* fighters, became part of Hawker Siddeley in 1920.

Handley Page, established in 1909, went out of business in 1970. In the preceding 61 years it made some heavy bombers for World War I and the WB airliner of 1920. This was followed by the HP42 airliner and some bombers for World War II, beginning with the *Hampden* and ending with the widely-used *Halifax.* Since then, the most notable aircraft was the *Victor,* which joined the *Valiant* and *Vulcan* to form the RAF's V-bomber force.

Fairey Aviation, which began in 1915, built the *Battle* light bomber, the *Swordfish* biplane torpedo bomber, and the *Rotodyne* VTOL airliner. It went into receivership in 1977. Gloster Aircraft was another casualty of the post-war shrinkage. In its early days it built the *Gladiator* biplane fighter that made a name for itself in Malta. Its *Meteor* was among the world's first jet fighters, and its final product was the *Javelin* twin-jet delta-wing fighter before closing down in 1956.

Some other well-known British companies such as Bristol, known for its *Blenheim* fighter and *Britannia* airliner; English Electric, known for its *Canberra* light jet bomber and its *Lightning* jet fighter, and Vickers with its *Spitfire, Wellington* bomber, *Viscount* turboprop airliner and *Valiant* heavy bomber, were absorbed into British Aircraft Corporation (BAC) in 1960, and this subsequently became part of BAE Systems.

So now we have BAE pretty well running the show in the U.K. apart from helicopters and some small enterprising companies that persevere with light aircraft. BAE is proving to be a competitive organization with some advanced products in a somewhat limited sphere. BAE is also a major partner with U.S. companies such as Lockheed Martin on the F-35. A project in which the U.K. has a vital interest—their new aircraft carrier, for example, is designed specifically to accommodate that aircraft.

In the United States, World War II companies faced a similar dilemma. Although some attempted to provide some marketable products, many eventually decided to leave the aircraft business. Remember the Brewster *Buffalo* and *Buccaneer?* This company started in the mid-1930s, and their *Buffalo* was sold to several countries. The first monoplane fighter in the U.S. Navy, but it and its follow-on were not as good as the competition, so Brewster airplanes disappeared from the scene in mid-1944.

Consolidated Aircraft's story is more complex. It designed and built the famous B-24 *Liberator* bomber and *Catalina* flying boat. In 1943, Vultee Aircraft took control, and the name changed to Consolidated

Vultee; then it changed again to Convair. In 1954, it finally became the Convair Division of General Dynamics Corporation. They produced such notable aircraft as the B-36 six-engined bomber, the F-102 and F-106 fighters, the B-58 supersonic bomber, the Convair 880, and 990 airliners, and the F-111 swing-wing fighter. They went out of business in 1996, and Lockheed took over their Fort Worth manufacturing facility and the F-16 fighter, which ultimately became an extremely successful aircraft as part of Lockheed Martin's stable.

The Curtiss Wright Corporation was the largest aircraft manufacturer in the U.S. at the end of World War II. Its products included the P-40 fighter and the C-46 *Commando* transport. Curtiss attempted a revival when the war ended, but it had not kept up with the latest technology. It was overcome by designs from companies such as Lockheed, North American, and Northrop, leading it to go out of the aircraft business in 1948.

Douglas Aircraft joined forces with McDonnell in 1967, resulting in a company well-entrenched company in both military and commercial activities—the F-15/F-18 fighters, C-17 cargo transport and the DC- 10 airliners. However, Boeing took over this in 1997, marking the end of a legendary company founded by Donald Douglas in 1921. It had produced the iconic DC-2/DC-3, some well-known fighters and dive bombers for World War II, and the whole series of DC airliners for worldwide operation. Like his contemporaries in the U.K., Alliott Verdon-Roe (Avro) and Geoffrey de Havilland, Douglas was one of the early aviation pioneers whose name became synonymous with a great company that ultimately followed the sands of time.

Fairchild was another old U.S. company no longer designing and building airplanes. Sherman Fairchild founded it in 1925, and its latest airplane was the A-10 *Warthog*, a ground attack aircraft that is currently being phased out of service.

North American Aviation's first major products were the T-6 *Harvard* trainer and the B-25 *Mitchell* bomber, but it was the P-51 *Mustang* that put the company into prominence. They followed that with the F-86 *Sabre* and F-100 *Supersabre*, both first-class fighters. Their last major effort was the XB-70 *Valkyrie* supersonic bomber that was abandoned after its flight test program, and in 1970 the company merged with Rockwell. This part of Rockwell then joined Boeing.

Republic P-47 fighters were used extensively to escort U.S. bombers in World War II. After the war, Republic produced the *Seabee* amphibian and the F-84 and F-105 fighters, after which in 1965 it became part of Fairchild. As noted above, that company is no longer in the aircraft business.

Ryan was notable in its early years for designing and building Lindbergh's *Spirit of St. Louis* and the PT-22 trainer. In the 1950s, it developed a lift-fan airplane and finally the XV-5B tail sitter fighter, similar in principle to Lockheed's XFV-1 *Pogo Stick* and Convair's XFY-1, all of which attempted to obtain a viable VTOL fighter. Subsequently, Ryan was taken over by Teledyne in 1968, and Teledyne Ryan later became part of Northrop Grumman.

So, in the U.S., there are now essentially three major aircraft companies— Boeing, Lockheed Martin and Northrop Grumman, plus a number of smaller companies such as Beech, Bell, Cessna, and Piper.

In Germany, think of famous World War II companies such as Dornier, Focke-Wulf, Junkers, and Messerschmitt. They were notable for the Do 17, Ju 86, Ju 88, He 111 bombers, the FW 190, Me 109, and Me 110 fighters, plus the Me 163 rocket-propelled interceptor. Dornier continued after the war, first in Spain and then back in Germany where they joined with Dassault to build the Alpha Jet trainer, and they produced the Do 328 small airliner before becoming part of the Daimler Benz.

Heinkel also continued after the war, building Lockheed F-104s until it was absorbed into the Messerschmitt-Bolkow-Blohm (MBB) company in 1980. Focke-Wulf merged with VFW, which later became part of the European Aeronautic Defence and Space Company (EADS).

Like the other German companies, Messerschmitt was not allowed to build aircraft for several years after the war and turned to non-aeronautical work to stay alive. Then in 1968, it merged with the Bolkow company and later with the aviation division of Blohm and Voss, resulting in a new company—Messerschmitt-Bolkow-Blohm.

While companies in the U.K. and U.S. have been closing and merging, France has had an interesting change In that country, the aircraft industry was reborn. At the end of World War II, it had no such industry, but today it is thriving. However, to a great extent, it is a combination of other European countries providing their expertise and manufacturing capability to a French base. In addition, the U.K. is also a major contributor to this enterprise. It all started with Nord Aviation and Sud Aviation in the 1950s. Their only significant airplane was the *Caravelle* from Sud. Then this company got together with British Aircraft Corporation (BAC) to begin work on the *Concorde* supersonic airliner.

Aerospatiale took over Sud and Nord and continued the partnership with BAC to develop the *Concorde*. Later, this company joined several European companies to form the EADS.

Airbus's story is more complicated due to the politics and proprietary rights of the several countries and companies involved. It began as a consortium of European and U.K. companies in France, Germany, Spain, and the U.K.. These companies had been developing designs to meet their own perceived needs, but production was economically impractical due to restricted markets. However, they recognized that by joining together, they could succeed, and their first project was the A300 twin-jet airliner. After a somewhat disappointing start, it emerged as a very marketable product, even penetrating the U.S. airlines.

Following this, the A320 was the world's first fly-by-wire airliner. To illustrate the multi-nation involvement, the wings were built in the U.K., the front and aft fuselage in Germany, the horizontal stabilizer in Spain, and the cockpit, controls, and total assembly were in France. This was followed by later aircraft in the 300-series, successfully competing with Boeing on a worldwide scale. Their latest version is the A380—the world's largest airliner.

Dassault is "the other" French company. It concentrated on two types of aircraft, fighters and business jets, which have been a success. The *Mirage* and *Rafale* fighters are relatively small, nimble aircraft that cost less than some of the more sophisticated fighters being marketed today. The *Falcon* business jet has emerged in various forms and the 8F version made its first flight in February 2015. With its size increase and longer range, it now competes with the Bombardier and Gulfstream biz jets.

There have also been two other consortiums formed in Europe. The first was Panavia, combining Aeritalia, BAE, and MBB from Italy, the U.K., and Germany. It produced the *Tornado* twin-jet multirole fighter in 1974 and is still in service. The second was Eurofighter, formed in 1986. It initially included companies from France, Italy, and the U.K. The French dropped out, but Germany and Spain were added. That company produced the *Typhoon*, a delta wing twin-jet fighter operational in 2008 and is in service with the German Air Force, Royal Air Force, Spanish Air Force, and Royal Saudi Air Force.

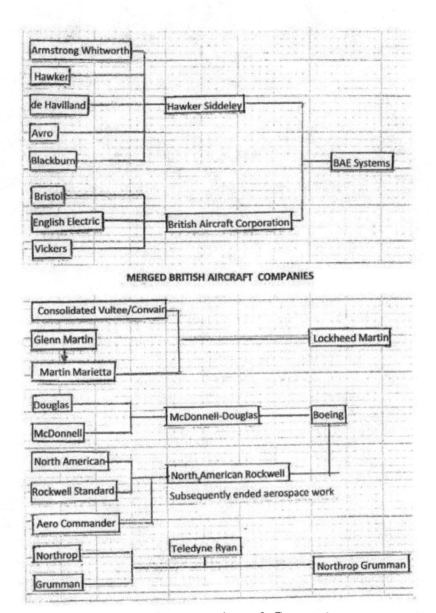

MERGED BRITISH AIRCRAFT COMPANIES

Merged American Aircraft Companies

CHAPTER 14
A NEW CENTURY OF AVIATION

In the last 100 yearsve, we've seen:

- "Ordinary" passengers flying internationally at 1500 mph.
- Airplanes carrying two main battle tanks.
- Airliner cabins accommodating hundreds of people across the oceans for vacations abroad.
- Reconnaissance airplanes flying at three times the speed of sound, at 15 miles altitude, taking super-accurate photos of what's below.
- Fighters can destroy a bomber with one missile while approaching each other at supersonic speeds at night.
- Bombs guided to hit targets precisely while the bombers are protected from chaff and flares, not guns.

It is doubtful that such advances would have been predicted when automobiles and planes were in their infancy, when tanks had just been invented, and when railroads were the most sophisticated form of travel. But the demands of two world wars, with associated funding, changed a lot of that, plus the fortuitous combination of the right people with the right skills, and in some cases, the exemplary courage.

We can now see where aviation is heading in the short term, but many factors influence longer term prediction. Nuclear weapons have proliferated so that many nations can quickly upset world peace, which could be caused by accident. Unfortunately, advancements in unmanned weapons can result in unintended consequences, and it is the recognition of this fact that will encourage the retention of manned vehicles for critical situations.

With the world in turmoil due to the perceived threats between major powers, religious wars or animosity, and continuous battles between various parts of Africa, national budgets are strained to keep up. Nevertheless, this turmoil is accelerating military inventiveness and, to a great extent, development in the commercial world.

Unmanned Aerial Vehicles (UAV), sometimes collectively referred to as drones, have become widespread. In Afghanistan and Pakistan, they were used to watch Osama bin Laden as he moved within his compound. In Yemen, they attacked and destroyed an automobile carrying terrorists, while on the other end of the scale, Amazon plans to use them to deliver packages, while others have become a hazard to airliners.

Although the history of UAVs is beyond the scope of this chapter, it is noteworthy that the first such vehicle was the World War II V-1 Buzz Bomb. It was rail-launched by the Germans against Great Britain, was powered by a ramjet and was not under control from the ground. Other UAVs, such as the Ryan *Firebee*, was in operation by the U.S. around 1957, and some classified projects were developed for the Vietnam War. But we had to wait until the Gulf War, and its successors before UAVs became widely and overtly used. They are now being used by the armed forces of more than 50 countries.

The General Atomics MQ-1 *Predator* was the most widely used UAV in the post-1995 period. The USAF and CIA used it for reconnaissance, and it was later equipped with *Hellfire* missiles to observe, track, and sometimes destroy terrorist activities. It can fly for over 400 miles, loiter for 14 hours, and return to base. It was followed in 2007 by the larger MQ-9 *Reaper*, also from General

Atomics. It had increased power, was faster, and had fifteen times more payload. Like its predecessor, it is controllable from the ground, often thousands of miles away via satellite link.

The Northrop Grumman RQ-4 *Global Hawk* is a relatively large UAV used for long-endurance, high-altitude, high-resolution surveillance. For this, it is equipped with long-range electro-optical sensors and was used by the USAF in Afghanistan, and by the U.S. Navy for surveillance over large areas of the ocean.

Boeing developed the *Condor* technology demonstrator in the late 1980s— a UAV that climbed to 67,000 feet in a test flight and could stay aloft for nearly 2 ½ days. After further work, it was abandoned. Boeing developed the *Scan Eagle*. A low-cost, long-endurance, 10 feet span UAV that is catapult-launched, autonomous, and has operated for several years with the U.S. Marine Corps, the U.S. Navy and a U.K. UAV program.

Lockheed Martin developed several UAVs, including the *Fury*, with 15 hours of endurance and the largest payload of any runway-independent vehicle. It is rail-launched and net-recoverable. The USAF operates its RQ-170 *Sentinel* flying wing to collect intelligence in Afghanistan. It shadowed Osama bin Laden and took movies of the Navy Seal operation that was watched in real-time by the U.S. President and his staff when OBL was killed. This UAV also operates from South Korea to keep an eye on the nefarious activities of its northern neighbor.

In contrast to these large, heavy, long-endurance UAVs is the AeroVironment RQ-11 *Raven*. A hand-held 40-pound "vehicle" that has a range of about six miles and speeds of about 60 mph. It is propeller- driven, using an electric motor, and went into production in 2006 for the U.S. Army, USAF, and Marine Corps. It is used to peek around the corner of buildings, look over walls and hills, and help in the street fighting that seems to be more and more prevalent these days. NASA has developed some ultra-thin solar panels stuck on its wing to increase range.

Proceeding to the latest developments, there is the General Atomics *Avenger*, a follow-on to the *Predator*, the Boeing X-15A to provide autonomous suppression of enemy air defenses, the Northrop Grumman X-47B which made the first UAV carrier landings in 2013, and a flying wing (mini B-2 shape) surveillance/strike UAV for the U.S. Navy that is planned to have an operating radius of 1200 nm. Northrop Grumman is developing the RQ-180 with Intelligence, Surveillance, Reconnaissance (ISR) capabilities, a classified program that incorporates the latest stealth technology. It will have a longer range or endurance than the *Reaper* and be in the same size category as the *Global Hawk*. In the U.K. and France, the BAE/Dassault Future Combat Air System is planned to be operational by 2030; the BAE Systems *Taranis*, a fighter-sized UAV that first flew in 2013, and the Dassault *Neuron* flew in 2012 to test sensors and stealth technology.

Lockheed Martin is active in cargo-carrying drones for both military and civil use. One is a pilotless helicopter, and the other, the *Indago*, is a small, hand-launched, slick-looking quadcopter. With military budgets tightening, LM hopes to market these to civil customers. It is small enough to be carried in the back seat of a car and would be suitable for the police to use in apprehending criminals. It could also be used for border patrol, pipeline monitoring or crop inspection. LM is also teamed with Kaman Aerospace to develop the K-MAX small unmanned helicopter for cargo carrying. It went one step further with *Ares*, a VTOL joint venture with Piasecki. It is powered by two rotatable ducted fans and can be either autonomous or controlled from the ground to carry payloads up to 3,000 pounds. As such, it is potentially very useful in delivering/extracting cargo or troops from very small areas without exposing the crew to enemy fire. Commercially, it would be well suited for fighting forest fires, pipeline exploration, agricultural or forestry activity, response to emergencies such as earthquakes, floods, and hurricanes, and the delivery of goods/services where access is limited or dangerous.

So for the first hundred years, we were teaching people how to fly airplanes, and in the second hundred, we're teaching airplanes to fly themselves! UAVs have already proved their worth, and the USAF is now training more people to fly them than to be conventional pilots. They save lives by watching enemy movements from a safe distance, attacking the enemy without human exposure, or observing natural disasters where volcanic gases or weather conditions could be hazardous. And like bomb disposal robots have been used for several years, they can be of great assistance in safely combating the terrorist or criminal elements in our midst.

In the years to come, UAVs will become a major part of an aeronautical activity in the search for maximum effectiveness, minimum cost, and the saving lives in both the civil and military spheres. We know that airplanes can be programmed to operate without human control. Still, it is an example of where humans must be left in the loop to recognize the fallibility of electrical and mechanical systems. A similar situation prevails in the military area where a human must be kept in the control of a bomber deploying an A-bomb to avoid the destruction of a city and the possible start of a war because of a communication error, for example. The simple joy of flying will never be replaced by a drone, so personal aircraft will continue to be produced, some capable of supersonic speeds.

One area that is being examined is the use of alternate fuels and the reduction of noise as well as atmospheric pollution by aircraft. Boeing, for example, developed the *Phantom Eye* UAV that flies at high altitudes, has long endurance, and is powered by liquid nitrogen. Its only byproduct is water! Its *Phantom Ray* is used to test new high technology, and will climb to 60,000 feet and stay up there for seven to ten days. Noise reduction over the airport and landing paths is an important issue for surrounding communities, and this has been addressed to some extent by modifying airplane operating procedures. However, it is still an area that needs work. As airplanes get bigger to satisfy demand, so do the engines. The Boeing 707-120, for example, weighed 207,000 pounds, while the 747-100B, ten years later, weighed three times as much (735,000 pounds). To use the same airfield lengths and achieve similar speeds, the engines on the 747 were four times more powerful. No matter how much high-tech is used, the 747 has to be noisier than its predecessor. Progress will continue in reducing engine noise, but some of the latest projects will increase engine noise while reducing fuel use and carbon emissions. This refers to the open-rotor engine.

Rolls-Royce and others are studying a gas turbine engine where contra- rotating propellers are mounted on the aft end of the engine's turbine section, the so-called open-rotor. Fuel savings as high as 30 percent are claimed due to the arrangement's high propulsive efficiency. However, it is noisy, and the space required for the propeller blades severely impacts installation possibilities. One solution is to mount the engine on pylons from the aft fuselage, which is good from the safety standpoint in that a detached blade would not penetrate the passenger cabin.

Electric power is an area that will continue to be explored in years to come and tentative steps have been made in this direction for low-payload, very long endurance flights. NASA has conducted many tests on electric-powered vehicles since the late 1990s. One such vehicle is the *Helius*—an all-wing plane with 247 feet span and five landing gears distributed along its span. It first flew in September 1999 and is powered by 14 electric motors, using a solar or cell system for a long-endurance mission. In one piloted flight, it reached an altitude of 96,803 feet, but its life ended when severe weather over the Pacific broke it up during a remotely-controlled flight.

Another interesting project is the Airbus E-fan, a small two-passenger airplane with lithium-ion batteries to power two electric engines. The engines drive two ducted variable-pitch fans, and the batteries last for an hour, after which they can be recharged in one hour. It flew in April 2014, and Airbus plans to produce it using Daher Socata to complete the design, and later they contemplate a regional airline version, the E-Thrust.

Several other companies, large and small (Boeing and Hugues Duval for example) are also studying electric power and fuel cells. NASA is in the early stages of examining a demonstrator, the Hybrid-Electric Integrated System Testbed (HEIST). It has a 31-feet span wing, mounted on a truck with 18 electric-powered engines driving small propellers. Its purpose is to investigate the effect of the airflow from these props as it passes over the wing, providing additional lift due to the high- speed flow over the leading edge. When used on a light airplane, one estimate is that this technique could allow the wing area to be reduced by as much as 60 to 70 percent. When the airplane is at a higher spped, these props could be folded with normal propulsion being provided by the conventional propellers. In Switzerland, two pilots, Bernard Picard and Andre Borschberg, embarked on an ambitious project—flying an electric-powered plane around the world. Called the *Solar Impulse 2, i*t has four 17.5 hp electric motors, weighs a mere 5,000 pounds, and has a 236-feet wing span. Its power is obtained from 1,400 lithium-ion batteries and solar cells. Some of the long-distance flight was in darkness when the batteries were used and were subsequently replenished from the solar cells when daylight returned. The pilots recognize that this is not likely to have commercial applications. Still, they are anxious to show how the world's energy consumption can be drastically reduced by using this technology, perhaps for UAVs used for transmission of telecommunications.

Aircraft structural design has always been a prime research, development, and advancement area. We started with wood and fabric and continued through the long series of aluminum alloys, with high-strength steels and titanium employed where applicable. In recent years carbon fiber and composite materials have been the favorite, and in years to come, they will be improved as new techniques, and material variations are employed. Although they have some problems, such as the expeditious repair of battle damage, they can be overcome, and their weight-saving plus lack-of-corrosion are predominant. A major research item is the so-called morphing structure that has a great impact on the airplane configuration and overall efficiency.

Birds have a morphing structure! They change their wing shape and size automatically to suit flight conditions and flight mode like soaring, takeoff, landing and so on. Leonardo da Vinci was probably the first to analyze a bird's anatomy, including its bone detail and joints, but he came up with the wrong conclusion. He thought flapping wings were necessary. Some early pioneers thought so, too, with disastrous results. It was not until Cayley and the Lilienthal brothers came along that bird flight was properly figured out. The effects of varying wing camber to change lift and the extending of pinion feathers increase wing area for takeoff and landing. The Wright Brothers emulated bird flight by having camber and by twisting their wings for roll control, and J.A.D. McCurdy had ailerons for roll.

Subsequently, rudders, elevators, flaps, ailerons, spoilers, and leading-edge slats were introduced for control and changing wing lift. But these came at a price, and designers are constantly faced with achieving an acceptable compromise between the varying demands of performance, control, and weight. The morphing aircraft attempts to change its configuration to suit a variety of missions. For example, a high-speed fighter could be adapted (morphed) in flight to make it suitable for a low-speed carrier landing. Reconfiguring can be accomplished by changing wing span, sweep, chord or camber, but all of these present formidable challenges in structural integrity, mechanism design, cost, and the minimization of weight.

Wing skins are typically used these days as the primary load-carrying element rather than the spars and ribs of older airplanes. Consequently, providing a flexible skin to change the camber or twist is difficult, and the linkages to do this can be complex. Various skin materials are being explored, such as reinforced shape memory polymer (SMP) that can be substantially stretched when heated and returns to its new shape when cooled and restored to its original shape when reheated. Another is a reinforced carbon fiber laminate that has internal electric heaters to do the same thing.

NASA has been involved in its "morphing project" for some time, using its Langley Research Center. That project develops and assesses the technologies and integrated component concepts to effectively enable multipoint adaptability. It joined Northrop Grumman in the "Smart Wing" program from the Defense Advanced Research Projects Agency (DARPA). It showed possibilities of moving seamless and unhinged control surfaces by 20 degrees up and down. Boeing led a similar program called "Smart Aircraft and Marine Project System Demonstration" (SAMPSON). DARPA has also been involved in the Next Generation Morphing Aircraft Structure (MAS) program to provide multi-role capabilities for upcoming military aircraft. Lockheed Martin worked on this and developed some radical designs that were tested and showed worthwhile possibilities. On a smaller scale, NASA is testing a Gulfstream bizjet that has FlexSys seamless control at the wing trailing edge to replace conventional flaps. Similar work is under way in Europe—the Smart Intelligent Aircraft Structures (SARISTU) and Smart Aircraft Morphing Technology (SMarph), for example. These are supplemented by studies underway at Bristol and Swansea Universities in the U.K. At Swansea, for example, they are examining a system to provide large changes in airfoil camber.

In summary, it is unlikely that morphing will be a major factor in commercial aircraft design. They are predominantly concerned with a simple mission—carrying passengers from A to B as quickly and cheaply as possible. But this new technology can have great potential for military aircraft, and initially for UAVs.

The bomber design will soon be at the forefront again, and Northrop Grumman won the contract for the U.S. Long-Range Strike Bomber (LRS-B). No published details are available on performance or payload, but it is expected to have a very high speed, intercontinental range and can accommodate nuclear weapons and the Massive Ordnance Penetrator (MOP) bunker-buster weapon. It also comes when the budgets are being strained in the western powers as they strive to participate in small wars in the Middle East, keep an eye on Russian expansion into neighboring countries, and recognize the growing Chinese threat in the Far East.

Commercial aircraft will be incrementally improved in the near term by incorporating research conducted in Europe's Clean Sky program, by NASA's Environmentally Responsible Aviation (ERA) program, and from studies by Airbus and Boeing. Laminar flow control is a prime example. Essentially a method for maintaining perfectly-attached airflow over the wing and tail surfaces, thereby increasing the efficiency of those parts. On a smaller scale, the fin and rudder size can be reduced substantially by using a series of small jets near the fin's leading edge to blow air over the surface with consequent reductions in drag and weight. One problem with producing laminar flow over the wings is that small holes in the leading edge that suck in the airflow to keep it attached become clogged with insects. Current research is now focused on providing a coating to prevent insect contamination.

NASA's ERA program also addresses noise reduction, fuel burn, overall drag, and emissions. Taken together, these are likely to have profound implications in the fairly near future, with less noise, lower carbon, and lower operating cost.

We can also expect to see radically new winglets for reduced drag, solar panels on the wings to provide power, and a host of nitpicking improvements here and there to improve aerodynamic efficiency and reduce drag.

Regarding overall aircraft configuration, it is likely that some merged wing-body designs will emerge. As the name implies, the fuselage has a flat oval cross-section. It blends smoothly into an enlarged wing root, a natural evolution that has not been used in previous years because of the circular section needed for pressurization. By clever design, this can be overcome particularly for unpressurized cargo-carrying.

General Atomics Predator

Northrop Grumman Global Hawk

Boeing Phantom Eye

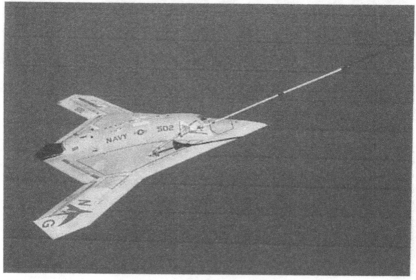

Northrop Grumman X-47D

NASA Study – Electric-Powered Aircraft

CHAPTER 15
EPILOGUE

During my 47 years in the aircraft industry, I have been fortunate enough to experience many aspects of flying. For the sheer joy and exuberance of flying, I've found it hard to beat the open-cockpit biplane *Tiger Moth*, doing the occasional maneuver – just for the fun of it! Going to the other extreme, I remember the luxury of flying first-class from Hong Kong to San Francisco in a Boeing 747. Involvement in test flying was always enjoyable. Sitting in the right seat of a *Mosquito* while the pilot was attempting stalling maneuvers, for example. And swinging across again to another extreme, there is the unusual for me and exciting experience of flying in a *Citation* biz jet to participate in a presentation or meeting at a subcontractors facility or an air force base. Naturally, this has to be balanced against the many hours in somewhat cramped accommodations, with lots of noise, little sleep and mediocre food as we sped across the Atlantic or Pacific Oceans. Still, even there, we had the exhilarating experience of viewing the tops of gorgeous clouds and, in one case, marvelling at the magnificent snow-covered mountains while approaching Anchorage, Alaska, from the south.

Amidst these travels, I have met engineers from Canada, China, France, Germany, Great Britain, South Korea, and the U.S. I observe a common bond between them that transcends international borders and political rivalry. It is encouraging and wonderful to openly discuss technology with Chinese, American, and British engineers sitting together. Naturally, participants avoid or deflect classified material, but this does not detract from the pleasant and valuable discussions. In Europe, we see British, German, and French engineers doing the same thing with no thought of the enmities that prevailed not too long ago.

To some extent, this is the inevitable result of aeronautical engineers migrating from one company and country to another. Still, I also look back to the exception to this in Russia. For this reason, only known to themselves, the engineers of that nation have not participated in this international togetherness. Perhaps it has been their unusual language and/or the secretive nature of their government.

In choosing subjects from my files for this book, I selected important, interesting, or significant subjects, so many fine projects were excluded for better or worse. Also, I have had minimal experience with general (light) aircraft or space flight, hence omitting such activities. Russian aircraft are not well represented because, even though there are many first-rate aircraft, we have not been aware of interesting or unusual stories in their development. A similar situation prevails in China at this time. Some 20 years ago, the Chinese were trying to develop an industry, but they have made rapid strides since then, and, since most of their products are military, our knowledge of their activities is limited.

Brazilian aircraft are not discussed in early chapters because their industry is relatively new, founded in 1969, and yet its Embraer small airliners and KC-390 transport are impressive. The latter will provide some competition for replacement sales of the venerable C-130. So far, it is in the flight test stage and capable of being a tanker and a transport. It can carry 64 airborne troops or six (68" x 108") pallets with a total payload of 52,029 pounds. Its range is 1,400 nm with a full payload.

The C-130J has a similar capacity but twice the range, and over the years, it has been adapted and proved in a wide variety of roles, so Embraer has a substantial challenge to match it. However, it has made a promising start, and we watch developments with interest.

Naturally, those of us in the industry always wonder what the advanced design departments are studying at the various companies. I suspect that Boeing is looking at supersonic or hypersonic bombers, and Lockheed Martin is also looking at alternatives in that arena; but what, if anything, is Lockheed Martin doing about a C-130 replacement?

Although it was out of my general sphere of activity, I was impressed by the Bell-Boeing V-22 *Osprey* multi-mission tilt-rotor aircraft. The concept wasn't new, of course, but it is the first aircraft of this type to be thoroughly successful. This must be attributed to the current availability of a highly-efficient propulsion system, lightweight composite materials, electronics, and engineers' skills. The wing tip-mounted engines can rotate to provide VTOL, STOL, and CTOL. It flies quite fast, has a nearly 400 nm combat radius, and can carry up to 32 troops or 20,000 pounds internal payload. The USAF and Marine Corps have successfully used it for several missions. It is eminently suitable for deploying a small force to save hostages, delivering humanitarian aid to devastated areas, and the inevitable deploying and retrieving of special forces in critical battlefield scenarios. In the U.K., Augusta Westland now Leonardo has produced the AW609 tiltrotor with a similar concept, and it is, at the time of writing, about to go into production.

Along these same lines, Lockheed Martin Skunk Works is leading a team with Piasecki to develop a later vehicle, the Ares, to perform a similar task but unmanned. It will have high speed and VTOL capability, using two large ducted fans that swivel to the required angles. It will be able to hover, and most importantly, the so-called fuselage ("mission module") is suspended beneath the flight module and can be easily changed to accommodate a variety of missions. Consequently, the basic element can be quickly changed for cargo carrying, casualty evacuation, intelligence surveillance, or reconnaissance. With no crew members exposed to enemy fire, it will be ideally suited for dangerous missions, and its small size will enable it to operate from small areas.

For another glimpse of the reasonably near future, examine the *Strategic Defence and Security Review (SDSR)* prepared in the U.K. At a lecture presented to the Royal Aeronautical Society ("*SDSR 2015,*" *Aero Space, March 2015)*, discussions revealed some of the concerns and possible solutions that must be addressed. The U.K. is of particular interest because its military requirements and capabilities are a highly condensed version of those in the U.S.—a worldwide presence to maintain peace and order with a somewhat limited budget and manpower availability. The speakers emphasized, for example, that the Lockheed Martin F-35 provides more capability than their *Tornado* and *Typhoon* aircraft. But the total number of frontline jet squadrons (the "combat mass") has decreased significantly, and one speaker noted that "no aircraft, however capable, can be in two places at once." One solution suggested that a mixed force of manned and unmanned aircraft may be affordable to increase that mass. However, remotely piloted air vehicles (RPAS) cannot perform all of the tasks undertaken by piloted aircraft, and their use may have ethical and legal implications. Another solution advocated the increased use of Tomahawk Land Attack Missiles (TLAM) fired from land and sea bases as a lower-cost system of expanding the capability of attacking a given target.

So, the U.K. is addressing the task of improving its combat forces, paying for a new first-class aircraft carrier, regaining its maritime patrol capability lost when the *Nimrod* was retired, and maintaining adequate manpower with inadequate funds. It is a situation that is repeated in many countries. But in Great Britain's case, the demands are more severe than most other countries because it is committed to maintaining a navy with world-ranking capability and an army that is regularly deployed to various parts of the world in mini-wars such as those in the Middle East and Afghanistan.

With all of the above considerations, where does this exciting adventure in aviation lead us now? Commercially, we can look forward to more people flying to more places at lower cost and greater comfort. The ghosts of Post, Earhart, and Lindbergh will see the local plumber packing his bags to skip over the Atlantic or Pacific to see the sights. Overall, this enhances international togetherness and diminishes conflict. However, on the military side, there is the uneasiness between Russia and "the West" and the emergence of China as a world player in the far east, with unknown intentions regarding the expansion of its borders. This inevitably leads western nations to explore hypersonic vehicles, highly sophisticated stealthy aircraft, and, to counter increasingly-effective anti- aircraft weapons, the use of remotely-guided or autonomous unmanned attack, reconnaissance, and surveillance vehicles. Then there is the constant threat of mini-wars and sectarian conflicts that must be suppressed so that they do not explode into a major conflagration. So, aviation has a big job to do, and we'd better do it right to save everyone!

Boeing V-22 Osprey

PICTURE CREDITS

Numbers in parentheses [] denote the pages at the end of which the picture is located

American Institute of Aeronautics & Astronautics – Landing Gears [p. 155-157] – from *Aircraft Landing Gear Design: Principles and Practices* by Noman S. Currey – with permission.

Airbus   A300 [p. 54], A340 [p. 54], A380 [p. 53], A400M [p. 38, 132,]

Avro Canada   Jetliner [p. 50]

Bill Larkin   Boeing 314 [p. 43]

Boeing   Boeing 737 [p. 51], Phantom Eye [p. 178]

Bombardier   Short Empire Flying Boat [p. 42]

De Havilland   Comet [p.122]

NASA   Lockheed Constellation [p. 114], Electric Powered a/c [p. 179]

Noman S. Currey – Spitfire [p. 100], Vampire [p. 46], CF-100 [p. 49], Wright Biplane [p. 17], Vega [p. 80], Maps [p. 82-83],

Mosquito [p. 93], Concorde [p. 148], Merged American Aircraft Companies [p. 165]

Royal Aeronautical Society – Cayley Inventions [p. 18], with permission from the National Aerospace Library/Mary Evans Picture Library.

U.S. Air Force Boeing B-17 [p. 44], P-38 [p. 45], F-86 [p. 47], F-80 [p. 48], Boeing 747 [p. 52], B-2 [p. 55], F-22 [p. 57], F-16 [p. 57], F-15 [p. 56], F-104G [p. 113], Lockheed XFV-1 [p. 111, 162], C-5 [p. 115], F-117 [p. 116], C-130J [p. 142], C-130 [p. 60]

Gunship [p. 140], C-130 At South Pole [p. 141], Predator [p. 176], Global Hawk [p. 177], X-47D [p. 178], Boeing V-22 [p. 185]

U.S. Marine Corps – Hawker Harrier [p. 59]

U.S. Navy   Curtiss NC-4 [p. 71]

BIBLIOGRAPHY

FOR GENERAL REFERENCE

Mondey, David, *The Complete Illustrated Encyclopedia of the World's Aircraft*, Chartwell Books Inc., Secaucus, NJ., 1978.

CHAPTER 1

Allen, J.E., *Aeronautics at Cayley's Second Century*, Aerospace, the Royal Aeronautical Society, London, Dec. 1973

Bonney, Walter F., *The Herritage of Kitty Hawk*, The Pegasus, Fairchild, Hagerstown, MD., April 1955

Cowlin, Dorothy, *The Flying Coachman of Brompton*, The Mercury, Scarborough, England, August 2,1986

Gibbs-Smith, Charles H., *Sir George Cayley, Father of Aerial Navigation, (1773–1857)*, Aeronautical Journal, Royal Aeronautical Society, London, April, 1974

Naylor, Derek, *A pioneer Who Was 100 Years Too Soon*, Yorkshire Evening Post, England, April 13, 1973

Parshall, Gerald, *Birds, Bicycles and Biplanes*, US News & World Report, New York, August 17/24, 1998

Pritchard, J, Laurence, *Sir George Cayley Bart., The Father of British Aeronautics*, Journal of the Royal Aeronautical Society, London, February, 1955.

Pritchard, J. Laurence, *The Wright Brothers and the Royal Aeronautical Society*, Journal of the Royal Aeronautical Society, London, December 1953

Read, Bill, *150 Years On–Cayley Glider Flies Again*, The Aerospace Professional, London, August, 2003

CHAPTER 3

Davy, M.J.B., *Aeronautics, Heavier-Than-Air Aircraft, Part 1 Historical Survey*, Science Museum, London, His Majesty's Stationery Office, 1949.

Heppenheimer, T.A., *Flight, One Hundred Years of Aviation in Photographs*, Carlton Books, London, 2003

Gilbert, James, *The Great Planes*, Grosset & Dunlap Inc., New York, 1970

CHAPTER 4

Blay, Roy, *A History of Lockheed*, Lockheed Horizons, Issue 12, Burbank, CA., 1983

Gilbert, James, *The Great Planes*, Grosset & Dunlap Inc., New York, 1970

Yenne, Bill, *Lockheed*, Bison Books, Greenwich, CT.,1987

CHAPTER 5

Birtles, Philip, *Mosquito*, Jane's, London, 1980

Birtles, Philip, *Planemakers:3 de Havilland,* Jane's, London, 1984

Clarkson, R.M., *Geoffrey de Havilland,* Journal of the Royal Aeronautical Society, London, February 1967

Ramsden, J.M., *First Composite Combat Aircraft,* Aerospace, London, November 1990

CHAPTER 6

Gilbert, James, *The Great Planes,* Grosset and Dunlap, New York, 1970

CHAPTER 7

Blay, Roy, *A History of Lockheed,* Grosset & Dunlap, New York, 1970

Rich, Ben R., *The Skunk Works Management Style – it's a Secret,* Aerospace, Royal Aeronautical Society, London, 1989

London, Sal, *First Lockheed Plane Flew 75 Years Ago,* Lockheed Southern Star, Marietta, GA., 1989

Slattery, Chad, *Secrets of the Skunk Works,* Air & Space, Smithsonian, Washington D.C., August, 2014

Wilson, Jim, *Flexible Flier,* Popular Mechanics, May, 2002

Yenna, Bill, *Lockheed,* Bison Book, Grenwich, CT., 1987

——, *Prospering in Tough Times,* Aviation Week & Space Technology, July 2014.

CHAPTER 8

Birtles, Philip J., *Planemakers:3, de Havilland,* Jane's Publishing Co., London, 1984

James, Brian, *Last Flight of the Trail-Blazing Comret*, Daily Mail, London, November 10, 1980

——-, *Comet Compendium*, Aeronautcs, London, June 1950

CHAPTER 9

Campagna, Palmiro, *Storms of Controversy*, Stoddart Publihing Co., Toronto, 1992

Zuuring, Peter, *The Arrow Scrapbook*, Arrow Alliance Press. Dalkeith, Ontario, 1999.

CHAPTER 10

Cefaratt, Gil., *Lockheed, The People Behind the Story*, Turner Publishing Co., Paducah, KY., 2002

Dabney, Joseph Earl, *Herc: Hero of the Skies*, Copple House Books, Lakemont, GA., 1979

Yenne, Bill, *Lockheed*, Bison Books, Bison Books Corp., Greenwich, CT., 1987

——-, *Days of Trial and Triumph,— a Pictorial History of Lockheed*, Lockheed Aircraft Corp., Burbank, CA., 1969

——-, *The Labours of Hercules*, Air International, UK., Nov., Dec., 1974

CHAPTER 12

Currey, Norman S., *Aircraft Landing Gear Design: Principles & Practices*, American Institute of Aeronautics & Astronautics, Washington DC, 1988

Currey, Norman S., *C-5 High Flotation Landing Gear*, Lockheed Georgia Company Quarterly, Marietta, GA., June 1968

CHAPTER 14

Dewer, Patrick, *Plane Speaking with Pat Dewer*, Aerospace, Royal Aeronautical Society, London, February 2015

Read, Bill, *Unmanned Future*, Aerospace, Royal Aeronautical Society, London, December 2014

Read, Bill, *Sun Flyer*, Aerospace, Royal Aeronautical Society, London, February 2015

Read, Bill, *Waiting in the Wings*, Aerospace, Royal Aeronautical Society, London, January 2015

Warwick, Graham, *Going the Duration*, Aviation Week & Space Technology, New York, February 2-15, 2015

Warwick, Graham, *Looking Ahead*, Aviation Week & Space Technology, New York, December 15-22, 2014

Warwick, Graham, *Electric Startup*, Aviatiion Week & Space Technology, New York, December 18, 2015

——, *The Sun Plane*, Popular Mechanics, March 2015

——, *Aerospace & Defense Intelligence from Essential Indiustry*, Aviation Week & Space Technology, New York, December 29, 2014- January 2015

——, *SDSR 2015*, Aerospace, Royal Aeronautical Society, London, March 2015